Deep Learning for the Life Sciences

Applying Deep Learning to Genomics, Microscopy, Drug Discovery, and More

Bharath Ramsundar, Peter Eastman,
Patrick Walters, and Vijay Pande

Beijing · Boston · Farnham · Sebastopol · Tokyo

Deep Learning for the Life Sciences

by Bharath Ramsundar, Peter Eastman, Patrick Walters, and Vijay Pande

Published by O'Reilly Media, Inc., 1005 Gravenstein Highway North, Sebastopol, CA 95472.

O'Reilly books may be purchased for educational, business, or sales promotional use. Online editions are also available for most titles (*http://oreilly.com*). For more information, contact our corporate/institutional sales department: 800-998-9938 or *corporate@oreilly.com*.

Development Editor: Nicole Tache
Acquisitions Editor: Mike Loukides
Production Editor: Katherine Tozer
Copyeditor: Rachel Head
Proofreader: Zachary Corleissen

Indexer: Ellen Troutman-Zaig
Interior Designer: David Futato
Cover Designer: Karen Montgomery
Illustrator: Rebecca Demarest

April 2019:   First Edition

Revision History for the First Edition
2019-03-27:   First Release
2021-11-12:   Second Release

See *http://bit.ly/deep-learning-life-science* for release details.

978-1-492-03983-9

[LSI]

Table of Contents

Preface

In recent years, life science and data science have converged. Advances in robotics and automation have enabled chemists and biologists to generate enormous amounts of data. Scientists today are capable of generating more data in a day than their predecessors 20 years ago could have generated in an entire career. This ability to rapidly generate data has also created a number of new scientific challenges. We are no longer in an era where data can be processed by loading it into a spreadsheet and making a couple of graphs. In order to distill scientific knowledge from these datasets, we must be able to identify and extract nonobvious relationships.

One technique that has emerged over the last few years as a powerful tool for identifying patterns and relationships in data is *deep learning*, a class of algorithms that have revolutionized approaches to problems such as image analysis, language translation, and speech recognition. Deep learning algorithms excel at identifying and exploiting patterns in large datasets. For these reasons, deep learning has broad applications across life science disciplines. This book provides an overview of how deep learning has been applied in a number of areas including genetics, drug discovery, and medical diagnosis. Many of the examples we describe are accompanied by code examples that provide a practical introduction to the methods and give the reader a starting point for future research and exploration.

Conventions Used in This Book

The following typographical conventions are used in this book:

Italic
> Indicates new terms, URLs, email addresses, filenames, and file extensions.

`Constant width`
> Used for program listings, as well as within paragraphs to refer to program elements such as variable or function names, databases, data types, environment variables, statements, and keywords.

`Constant width bold`
> Shows commands or other text that should be typed literally by the user.

`Constant width italic`
> Shows text that should be replaced with user-supplied values or by values determined by context.

 This element signifies a tip or suggestion.

 This element signifies a general note.

 This element indicates a warning or caution.

Using Code Examples

Supplemental material (code examples, exercises, etc.) is available for download at *https://github.com/deepchem/DeepLearningLifeSciences*.

This book is here to help you get your job done. In general, if example code is offered with this book, you may use it in your programs and documentation. You do not need to contact us for permission unless you're reproducing a significant portion of the code. For example, writing a program that uses several chunks of code from this book does not require permission. Selling or distributing a CD-ROM of examples from O'Reilly books does require permission. Answering a question by citing this book and quoting example code does not require permission. Incorporating a significant amount of example code from this book into your product's documentation does require permission.

We appreciate, but do not require, attribution. An attribution usually includes the title, author, publisher, and ISBN. For example: "*Deep Learning for the Life Sciences* by Bharath Ramsundar, Peter Eastman, Patrick Walters, and Vijay Pande (O'Reilly). Copyright 2019 Bharath Ramsundar, Karl Leswing, Peter Eastman, and Vijay Pande, 978-1-492-03983-9."

If you feel your use of code examples falls outside fair use or the permission given above, feel free to contact us at *permissions@oreilly.com*.

O'Reilly Online Learning

 For almost 40 years, *O'Reilly* has provided technology and business training, knowledge, and insight to help companies succeed.

Our unique network of experts and innovators share their knowledge and expertise through books, articles, conferences, and our online learning platform. O'Reilly's online learning platform gives you on-demand access to live training courses, in-depth learning paths, interactive coding environments, and a vast collection of text and video from O'Reilly and 200+ other publishers. For more information, please visit *http://oreilly.com*.

How to Contact Us

Please address comments and questions concerning this book to the publisher:

O'Reilly Media, Inc.
1005 Gravenstein Highway North
Sebastopol, CA 95472
800-998-9938 (in the United States or Canada)
707-829-0515 (international or local)
707-829-0104 (fax)

We have a web page for this book, where we list errata, examples, and any additional information. You can access this page at *http://bit.ly/deep-lrng-for-life-science*.

To comment or ask technical questions about this book, send email to *bookquestions@oreilly.com*.

For more information about our books, courses, conferences, and news, see our website at *http://www.oreilly.com*.

Find us on Facebook: *http://facebook.com/oreilly*

Follow us on Twitter: *http://twitter.com/oreillymedia*

Watch us on YouTube: *http://www.youtube.com/oreillymedia*

Acknowledgments

We would like to thank Nicole Tache, our editor at O'Reilly, as well as the tech reviewers and beta reviewers for their valuable contributions to the book. In addtion, we would like to thank Karl Leswing and Zhenqin (Michael) Wu for their contributions to the code and Johnny Israeli for valuable advice on the genomics chapter.

Bharath would like to thank his family for their support and encouragement during many long weekends and nights working on this book.

Peter would like to thank his wife for her constant support, as well as the many colleagues from whom he has learned so much about machine learning.

Pat would like to thank his wife Andrea, and his daughters Alee and Maddy, for their love and support. He would also like to acknowledge past and present colleagues at Vertex Pharmaceuticals and Relay Therapeutics, from whom he has learned so much.

Finally, we want to thank the DeepChem open source community for their encouragement and support throughout this project.

## CHAPTER 1

Why Life Science?

While there are many directions that those with a technical inclination and a passion for data can pursue, few areas can match the fundamental impact of biomedical research. The advent of modern medicine has fundamentally changed the nature of human existence. Over the last 20 years, we have seen innovations that have transformed the lives of countless individuals. When it first appeared in 1981, HIV/AIDS was a largely fatal disease. Continued development of antiretroviral therapies has dramatically extended the life expectancy for patients in the developed world. Other diseases, such as hepatitis C, which was considered largely untreatable a decade ago, can now be cured. Advances in genetics are enabling the identification and, hopefully soon, the treatment of a wide array of diseases. Innovations in diagnostics and instrumentation have enabled physicians to specifically identify and target disease in the human body. Many of these breakthroughs have benefited from and will continue to be advanced by computational methods.

Why Deep Learning?

Machine learning algorithms are now a key component of everything from online shopping to social media. Teams of computer scientists are developing algorithms that enable digital assistants such as the Amazon Echo or Google Home to understand speech. Advances in machine learning have enabled routine on-the-fly translation of web pages between spoken languages. In addition to machine learning's impact on everyday life, it has impacted many areas of the physical and life sciences. Algorithms are being applied to everything from the detection of new galaxies from telescope images to the classification of subatomic interactions at the Large Hadron Collider.

One of the drivers of these technological advances has been the development of a class of machine learning methods known as deep neural networks. While the technological underpinnings of artificial neural networks were developed in the 1950s and refined in the 1980s, the true power of the technique wasn't fully realized until advances in computer hardware became available over the last 10 years. We will provide a more complete overview of deep neural networks in the next chapter, but it is important to acknowledge some of the advances that have occurred through the application of deep learning:

- Many of the developments in speech recognition that have become ubiquitous in cell phones, computers, televisions, and other internet-connected devices have been driven by deep learning.

- Image recognition is a key component of self-driving cars, internet search, and other applications. Many of the same developments in deep learning that drove consumer applications are now being used in biomedical research, for example, to classify tumor cells into different types.

- Recommender systems have become a key component of the online experience. Companies like Amazon use deep learning to drive their "customers who bought this also bought" approach to encouraging additional purchases. Netflix uses a similar approach to recommend movies that an individual may want to watch. Many of the ideas behind these recommender systems are being used to identify new molecules that may provide starting points for drug discovery efforts.

- Language translation was once the domain of very complex rule-based systems. Over the last few years, systems driven by deep learning have outperformed systems that had undergone years of manual curation. Many of the same ideas are now being used to extract concepts from the scientific literature and alert scientists to journal articles that they may have missed.

These are just a few of the innovations that have come about through the application of deep learning methods. We are at an interesting time when we have a convergence of widely available scientific data and methods for processing that data. Those with the ability to combine data with new methods for learning from patterns in that data can make significant scientific advances.

Contemporary Life Science Is About Data

As mentioned previously, the fundamental nature of life science has changed. The availability of robotics and miniaturized experiments has brought about dramatic increases in the amount of experimental data that can be generated. In the 1980s a biologist would perform a single experiment and generate a single result. This sort of data could typically be manipulated by hand with the possible assistance of a pocket calculator. If we fast-forward to today's biology, we have instrumentation that is capable of generating millions of experimental data points in a day or two. Experiments like gene sequencing, which can generate huge datasets, have become inexpensive and routine.

The advances in gene sequencing have led to the construction of databases that link an individual's genetic code to a multitude of health-related outcomes, including diabetes, cancer, and genetic diseases such as cystic fibrosis. By using computational techniques to analyze and mine this data, scientists are developing an understanding of the causes of these diseases and using this understanding to develop new treatments.

Disciplines that once relied primarily on human observation are now utilizing datasets that simply could not be analyzed manually. Machine learning is now routinely used to classify images of cells. The output of these machine learning models is used to identify and classify cancerous tumors and to evaluate the effects of potential disease treatments.

Advances in experimental techniques have led to the development of several databases that catalog the structures of chemicals and the effects that these chemicals have on a wide range of biological processes or activities. These structure–activity relationships (SARs) form the basis of a field known as chemical informatics, or *cheminformatics*. Scientists mine these large datasets and use the data to build predictive models that will drive the next generation of drug development.

With these large amounts of data comes a need for a new breed of scientist who is comfortable in both the scientific and computational domains. Those with these hybrid capabilities have the potential to unlock structure and trends in large datasets and to make the scientific discoveries of tomorrow.

What Will You Learn?

In the first few chapters of this book, we provide an overview of deep learning and how it can be applied in the life sciences. We begin with machine learning, which has been defined as "the science (and art) of programming computers so that they can learn from data."[1]

Chapter 2 provides a brief introduction to deep learning. We begin with an example of how this type of machine learning can be used to perform a simple task like linear regression, and progress to more sophisticated models that are commonly used to solve real-world problems in the life sciences. Machine learning typically proceeds by initially splitting a dataset into a training set that is used to generate a model and a test set that is used to assess the performance of the model. In Chapter 2 we discuss

---

1 Furbush, James. "Machine Learning: A Quick and Simple Definition." *https://www.oreilly.com/ideas/machine-learning-a-quick-and-simple-definition*. 2018.

some of the details surrounding the training and validation of predictive models. Once a model has been generated, its performance can typically be optimized by varying a number of characteristics known as *hyperparameters*. The chapter provides an overview of this process. Deep learning is not a single technique, but a set of related methods. Chapter 2 concludes with an introduction to a few of the most important deep learning variants.

In Chapter 3, we introduce DeepChem, an open source programming library that has been specifically designed to simplify the creation of deep learning models for a variety of life science applications. After providing an overview of DeepChem, we introduce our first programming example, which demonstrates how the DeepChem library can be used to generate a model for predicting the toxicity of molecules. In a second programming example, we show how DeepChem can be used to classify images, a common task in modern biology. As briefly mentioned earlier, deep learning is used in a variety of imaging applications, ranging from cancer diagnosis to the detection of glaucoma. This discussion of specific applications then motivates an explanation of some of the inner workings of deep learning methods.

Chapter 4 provides an overview of how machine learning can be applied to molecules. We begin by introducing molecules, the building blocks of everything around us. Although molecules can be considered analogous to building blocks, they are not rigid. Molecules are flexible and exhibit dynamic behavior. In order to characterize molecules using a computational method like deep learning, we need to find a way to represent molecules in a computer. These encodings can be thought of as similar to the way in which an image can be represented as a set of pixels. In the second half of Chapter 4, we describe a number of ways that molecules can be represented and how these representations can be used to build deep learning models.

Chapter 5 provides an introduction to the field of biophysics, which applies the laws of physics to biological phenomena. We start with a discussion of proteins, the molecular machines that make life possible. A key component of predicting the effects of drugs on the body is understanding their interactions with proteins. In order to understand these effects, we begin with an overview of how proteins are constructed and how protein structures differ. Proteins are entities whose 3D structure dictates their biological function. For a machine learning model to predict the impact of a drug molecule on a protein's function, we need to represent that 3D structure in a form that can be processed by a machine learning program. In the second half of Chapter 5, we explore a number of ways that protein structures can be represented. With this knowledge in hand, we then review another code example where we use deep learning to predict the degree to which a drug molecule will interact with a protein.

Genetics has become a key component of contemporary medicine. The genetic sequencing of tumors has enabled the personalized treatment of cancer and has the potential to revolutionize medicine. Gene sequencing, which used to be a complex process requiring huge investments, has now become commonplace and can be routinely carried out. We have even reached the point where dog owners can get inexpensive genetic tests to determine their pets' lineage. In Chapter 6, we provide an overview of genetics and genomics, beginning with an introduction to DNA and RNA, the templates that are used to produce proteins. Recent discoveries have revealed that the interactions of DNA and RNA are much more complex than originally believed. In the second half of Chapter 6, we present several code examples that demonstrate how deep learning can be used to predict a number of factors that influence the interactions of DNA and RNA.

Earlier in this chapter, we alluded to the many advances that have come about through the application of deep learning to the analysis of biological and medical images. Many of the phenomena studied in these experiments are too small to be observed by the human eye. In order to obtain the images used with deep learning methods, we need to utilize a microscope. Chapter 7 provides an overview of microscopy in its myriad forms, ranging from the simple light microscope we all used in school to sophisticated instruments that are capable of obtaining images at atomic resolution. This chapter also covers some of the limitations of current approaches, and provides information on the experimental pipelines used to obtain the images that drive deep learning models.

One area that offers tremendous promise is the application of deep learning to medical diagnosis. Medicine is incredibly complex, and no physician can personally embody all of the available medical knowledge. In an ideal situation, a machine learning model could digest the medical literature and aid medical professionals in making diagnoses. While we have yet to reach this point, a number of positive steps have been made. Chapter 8 begins with a history of machine learning methods for medical diagnosis and charts the transition from hand-encoded rules to statistical analysis of medical outcomes. As with many of the topics we've discussed, a key component is representing medical information in a format that can be processed by a machine learning program. In this chapter, we provide an introduction to electronic health records and some of the issues surrounding these records. In many cases, medical images can be very complex and the analysis and interpretation of these images can be difficult for even skilled human specialists. In these cases, deep learning can augment the skills of a human analyst by classifying images and identifying key features. Chapter 8 concludes with a number of examples of how deep learning is used to analyze medical images from a variety of areas.

As we mentioned earlier, machine learning is becoming a key component of drug discovery efforts. Scientists use deep learning models to evaluate the interactions between drug molecules and proteins. These interactions can elicit a biological response that has a therapeutic impact on a patient. The models we've discussed so far are *discriminative models*. Given a set of characteristics of a molecule, the model generates a prediction of some property. These predictions require an input molecule, which may be derived from a large database of available molecules or may come from the imagination of a scientist. What if, rather than relying on what currently exists, or what we can imagine, we had a computer program that could "invent" new molecules? Chapter 9 presents a type of deep learning program called a *generative model*. A generative model is initially trained on a set of existing molecules, then used to generate new molecules. The deep learning program that generates these molecules can also be influenced by other models that predict the activity of the new molecules.

Up to now, we have discussed deep learning models as "black boxes." We present the model with a set of input data and the model generates a prediction, with no explanation of how or why the prediction was generated. This type of prediction can be less than optimal in many situations. If we have a deep learning model for medical diagnosis, we often need to understand the reasoning behind the diagnosis. An explanation of the reasons for the diagnosis will provide a physician with more confidence in the prediction and may also influence treatment decisions. One historic drawback to deep learning has been the fact that the models, while often reliable, can be difficult to interpret. A number of techniques are currently being developed to enable users to better understand the factors that led to a prediction. Chapter 10 provides an overview of some of these techniques used to enable human understanding of model predictions. Another important aspect of predictive models is the accuracy of a model's predictions. An understanding of a model's accuracy can help us determine how much to rely on that model. Given that machine learning can be used to potentially make life-saving diagnoses, an understanding of model accuracy is critical. The final section of Chapter 10 provides an overview of some of the techniques that can be used to assess the accuracy of model predictions.

In Chapter 11 we present a real-world case study using DeepChem. In this example, we use a technique called virtual screening to identify potential starting points for the discovery of new drugs. Drug discovery is a complex process that often begins with a technique known as *screening*. Screening is used to identify molecules that can be optimized to eventually generate drugs. Screening can be carried out experimentally, where millions of molecules are tested in miniaturized biological tests known as assays, or in a computer using virtual screening. In virtual screening, a set of known drugs or other biologically active molecules is used to train a machine learning model. This machine learning model is then used to predict the activity of a large set of molecules. Because of the speed of machine learning methods, hundreds of millions of molecules can typically be processed in a few days of computer time.

The final chapter of the book examines the current impact and future potential of deep learning in the life sciences. A number of challenges for current efforts, including the availability and quality of datasets, are discussed. We also highlight opportunities and potential pitfalls in a number of other areas including diagnostics, personalized medicine, pharmaceutical development, and biology research.

## CHAPTER 2

Introduction to Deep Learning

The goal of this chapter is to introduce the basic principles of deep learning. If you already have lots of experience with deep learning, you should feel free to skim this chapter and then go on to the next. If you have less experience, you should study this chapter carefully as the material it covers will be essential to understanding the rest of the book.

In most of the problems we will discuss, our task will be to create a mathematical function:

$$\mathbf{y} = f(\mathbf{x})$$

Notice that $\mathbf{x}$ and $\mathbf{y}$ are written in bold. This indicates they are vectors. The function might take many numbers as input, perhaps thousands or even millions, and it might produce many numbers as outputs. Here are some examples of functions you might want to create:

- $\mathbf{x}$ contains the colors of all the pixels in an image. $f(\mathbf{x})$ should equal 1 if the image contains a cat and 0 if it does not.
- The same as above, except $f(\mathbf{x})$ should be a vector of numbers. The first element indicates whether the image contains a cat, the second whether it contains a dog, the third whether it contains an airplane, and so on for thousands of types of objects.
- $\mathbf{x}$ contains the DNA sequence for a chromosome. $\mathbf{y}$ should be a vector whose length equals the number of bases in the chromosome. Each element should equal 1 if that base is part of a region that codes for a protein, or 0 if not.
- $\mathbf{x}$ describes the structure of a molecule. (We will discuss various ways of representing molecules in later chapters.) $\mathbf{y}$ should be a vector where each element

describes some physical property of the molecule: how easily it dissolves in water, how strongly it binds to some other molecule, and so on.

As you can see, $f(\mathbf{x})$ could be a very, very complicated function! It usually takes a long vector as input and tries to extract information from it that is not at all obvious just from looking at the input numbers.

The traditional approach to solving this problem is to design a function by hand. You would start by analyzing the problem. What patterns of pixels tend to indicate the presence of a cat? What patterns of DNA tend to distinguish coding regions from noncoding ones? You would write computer code to recognize particular types of features, then try to identify combinations of features that reliably produce the result you want. This process is slow and labor-intensive, and depends heavily on the expertise of the person carrying it out.

Machine learning takes a totally different approach. Instead of designing a function by hand, you allow the computer to learn its own function based on data. You collect thousands or millions of images, each labeled to indicate whether it includes a cat. You present all of this training data to the computer, and let it search for a function that is consistently close to 1 for the images with cats and close to 0 for the ones without.

What does it mean to "let the computer search for a function"? Generally speaking, you create a *model* that defines some large class of functions. The model includes *parameters*, variables that can take on any value. By choosing the values of the parameters, you select a particular function out of all the many functions in the class defined by the model. The computer's job is to select values for the parameters. It tries to find values such that, when your training data is used as input, the output is as close as possible to the corresponding targets.

Linear Models

One of the simplest models you might consider trying is a linear model:

$$\mathbf{y} = \mathbf{Mx} + \mathbf{b}$$

In this equation, $\mathbf{M}$ is a matrix (sometimes referred to as the "weights") and $\mathbf{b}$ is a vector (referred to as the "biases"). Their sizes are determined by the numbers of input and output values. If $\mathbf{x}$ has length T and you want $\mathbf{y}$ to have length S, then $\mathbf{M}$ will be an S $\times$ T matrix and $\mathbf{b}$ will be a vector of length S. Together, they make up the parameters of the model. This equation simply says that each output component is a linear combination of the input components. By setting the parameters ($\mathbf{M}$ and $\mathbf{b}$), you can choose any linear combination you want for each component.

This was one of the very earliest machine learning models. It was introduced back in 1957 and was called a *perceptron*. The name is an amazing piece of marketing: it has a science fiction sound to it and seems to promise wonderful things, when in fact it is nothing more than a linear transform. In any case, the name has managed to stick for more than half a century.

The linear model is very easy to formulate in a completely generic way. It has exactly the same form no matter what problem you apply it to. The only differences between linear models are the lengths of the input and output vectors. From there, it is just a matter of choosing the parameter values, which can be done in a straightforward way with generic algorithms. That is exactly what we want for machine learning: a model and algorithms that are independent of what problem you are trying to solve. Just provide the training data, and parameters are automatically determined that transform the generic model into a function that solves your problem.

Unfortunately, linear models are also very limited. As demonstrated in Figure 2-1, a linear model (in one dimension, that means a straight line) simply cannot fit most real datasets. The problem becomes even worse when you move to very high-dimensional data. No linear combination of pixel values in an image will reliably identify whether the image contains a cat. The task requires a much more complicated nonlinear model. In fact, any model that solves that problem will necessarily be *very* complicated and *very* nonlinear. But how can we formulate it in a generic way? The space of all possible nonlinear functions is infinitely complex. How can we define a model such that, just by choosing values of parameters, we can create almost any nonlinear function we are ever likely to want?

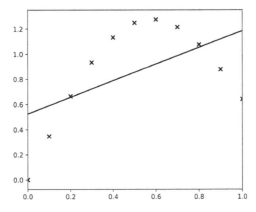

Figure 2-1. A linear model cannot fit data points that follow a curve. This requires a nonlinear model.

Multilayer Perceptrons

A simple approach is to stack multiple linear transforms, one after another. For example, we could write:

$$\mathbf{y} = \mathbf{M}_2 \varphi(\mathbf{M}_1 \mathbf{x} + \mathbf{b}_1) + \mathbf{b}_2$$

Look carefully at what we have done here. We start with an ordinary linear transform, $\mathbf{M}_1 \mathbf{x} + \mathbf{b}_1$. We then pass the result through a nonlinear function $\varphi(x)$, and then apply a second linear transform to the result. The function $\varphi(x)$, which is known as the *activation function*, is an essential part of what makes this work. Without it, the model would still be linear, and no more powerful than the previous one. A linear combination of linear combinations is itself nothing more than a linear combination of the original inputs! By inserting a nonlinearity, we enable the model to learn a much wider range of functions.

We don't need to stop at two linear transforms. We can stack as many as we want on top of each other:

$$\mathbf{h}_1 = \varphi_1(\mathbf{M}_1 \mathbf{x} + \mathbf{b}_1)$$

$$\mathbf{h}_2 = \varphi_2(\mathbf{M}_2 \mathbf{h}_1 + \mathbf{b}_2)$$

$$\ldots$$

$$\mathbf{h}_{n-1} = \varphi_{n-1}(\mathbf{M}_{n-1} \mathbf{h}_{n-2} + \mathbf{b}_{n-1})$$

$$\mathbf{y} = \varphi_n(\mathbf{M}_n \mathbf{h}_{n-1} + \mathbf{b}_n)$$

This model is called a *multilayer perceptron*, or MLP for short. The middle steps h_i are called *hidden layers*. The name refers to the fact that they are neither inputs nor outputs, just intermediate values used in the process of calculating the result. Also notice that we have added a subscript to each $\varphi(x)$. This indicates that different layers might use different nonlinearities.

You can visualize this calculation as a stack of layers, as shown in Figure 2-2. Each layer corresponds to a linear transformation followed by a nonlinearity. Information flows from one layer to another, the output of one layer becoming the input to the next. Each layer has its own set of parameters that determine how its output is calculated from its input.

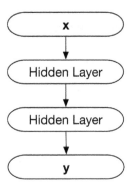

Figure 2-2. A multilayer perceptron, viewed as a stack of layers with information flowing from one layer to the next.

Multilayer perceptrons and their variants are also sometimes called *neural networks*. The name reflects the parallels between machine learning and neurobiology. A biological neuron connects to many other neurons. It receives signals from them, adds the signals together, and then sends out its own signals based on the result. As a very rough approximation, you can think of MLPs as working the same way as the neurons in your brain!

What should the activation function $\varphi(\mathbf{x})$ be? The surprising answer is that it mostly doesn't matter. Of course, that is not entirely true. It obviously does matter, but not as much as you might expect. Nearly any reasonable function (monotonic, reasonably smooth) can work. Lots of different functions have been tried over the years, and although some work better than others, nearly all of them can produce decent results.

The most popular activation function today is probably the *rectified linear unit* (ReLU), $\varphi(x) = \max(0, x)$. If you aren't sure what function to use, this is probably a good default. Other common choices include the *hyperbolic tangent*, tanh(x), and the *logistic sigmoid*, $\varphi(x) = 1/(1 + e^{-x})$. All of these functions are shown in Figure 2-3.

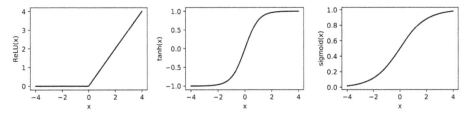

Figure 2-3. Three common activation functions: the rectified linear unit, hyperbolic tangent, and logistic sigmoid.

We also must choose two other properties for an MLP: its *width* and its *depth*. With the simple linear model, we had no choices to make. Given the lengths of **x** and **y**, the sizes of **M** and **b** were completely determined. Not so with hidden layers. Width refers to the size of the hidden layers. We can choose each $\mathbf{h}_i$ to have any length we want. Depending on the problem, you might want them to be much larger or much smaller than the input and output vectors.

Depth refers to the number of layers in the model. A model with only one hidden layer is described as *shallow*. A model with many hidden layers is described as *deep*. This is, in fact, the origin of the term "deep learning"; it simply means "machine learning using models with lots of layers."

Choosing the number and widths of layers in your model involves as much art as science. Or, to put it more formally, "This is still an active field of research." Often it just comes down to trying lots of combinations and seeing what works. There are a few principles that may provide guidance, however, or at least help you understand your results in hindsight:

1. An MLP with one hidden layer is a *universal approximator*.

   This means it can approximate any function at all (within certain fairly reasonable limits). In a sense, you never need more than one hidden layer. That is already enough to reproduce any function you are ever likely to want. Unfortunately, this result comes with a major caveat: the accuracy of the approximation depends on the width of the hidden layer, and you may need a very wide layer to get sufficient accuracy for a given problem. This brings us to the second principle.

2. Deep models tend to require fewer parameters than shallow ones.

   This statement is intentionally somewhat vague. More rigorous statements can be proven for particular special cases, but it does still apply as a general guideline. Here is perhaps a better way of stating it: every problem requires a model with a certain depth to efficiently achieve acceptable accuracy. At shallower depths, the required widths of the layers (and hence the total number of parameters) increase rapidly. This makes it sound like you should always prefer deep models over shallow ones. Unfortunately, it is partly contradicted by the third principle.

3. Deep models tend to be harder to train than shallow ones.

   Until about 2007, most machine learning models were shallow. The theoretical advantages of deep models were known, but researchers were usually unsuccessful at training them. Since then, a series of advances has gradually improved the usefulness of deep models. These include better training algorithms, new types of models that are easier to train, and of course faster computers combined with larger datasets on which to train the models. These advances gave rise to "deep learning" as a field. Yet despite the improvements, the general principle remains true: deeper models tend to be harder to train than shallower ones.

Training Models

This brings us to the next subject: just how do we train a model anyway? MLPs provide us with a (mostly) generic model that can be used for any problem. (We will discuss other, more specialized types of models a little later.) Now we want a similarly generic algorithm to find the optimal values of the model's parameters for a given problem. How do we do that?

The first thing you need, of course, is a collection of data to train it on. This dataset is known as the *training set*. It should consist of a large number of (**x,y**) pairs, also known as *samples*. Each sample specifies an input to the model, and what you want the model's output to be when given that input. For example, the training set could be a collection of images, along with labels indicating whether or not each image contains a cat.

Next you need to define a loss function $L(\mathbf{y}, \widehat{\mathbf{y}})$, where **y** is the actual output from the model and $\widehat{\mathbf{y}}$ is the target value specified in the training set. This is how you measure whether the model is doing a good job of reproducing the training data. It is then averaged over every sample in the training set:

$$\text{average loss} = \frac{1}{N} \sum_{i=1}^{N} L(\mathbf{y}_i, \widehat{\mathbf{y}}_i)$$

$L(\mathbf{y}, \widehat{\mathbf{y}})$ should be small when its arguments are close together and large when they are far apart. In other words, we take every sample in the training set, try using each one as an input to the model, and see how close the output is to the target value. Then we average this over the whole training set.

An appropriate loss function needs to be chosen for each problem. A common choice is the Euclidean distance (also known as the L_2 distance), $L(\mathbf{y}, \widehat{\mathbf{y}}) = \sqrt{\sum_i (y_i - \widehat{y}_i)^2}$. (In this expression, y_i means the i'th component of the vector **y**.) When **y** represents a probability distribution, a popular choice is the cross entropy, $L(\mathbf{y}, \widehat{\mathbf{y}}) = -\sum_i y_i \log \widehat{y}_i$. Other choices are also possible, and there is no universal "best" choice. It depends on the details of your problem.

Now that we have a way to measure how well the model works, we need a way to improve it. We want to search for the parameter values that minimize the average loss over the training set. There are many ways to do this, but most work in deep learning uses some variant of the *gradient descent* algorithm. Let θ represent the set of all parameters in the model. Gradient descent involves taking a series of small steps:

$$\theta \leftarrow \theta - \epsilon \frac{\partial}{\partial \theta} \langle L \rangle$$

where $\langle L \rangle$ is the average loss over the training set. Each step moves a tiny distance in the "downhill" direction. It changes each of the model's parameters by a little bit, with the goal of causing the average loss to decrease. If all the stars align and the phase of the moon is just right, this will eventually produce parameters that do a good job of solving your problem. ϵ is called the *learning rate*, and it determines how much the parameters change on each step. It needs to be chosen very carefully: too small a value will cause learning to be very slow, while too large a value will prevent the algorithm from learning at all.

This algorithm really does work, but it has a serious problem. For every step of gradient descent, we need to loop over every sample in the training set. That means the time required to train the model is proportional to the size of the training set! Suppose that you have one million samples in the training set, that computing the gradient of the loss for one sample requires one million operations, and that it takes one million steps to find a good model. (All of these numbers are fairly typical of real deep learning applications.) Training will then require *one quintillion* operations. That takes quite a long time, even on a fast computer.

Fortunately, there is a better solution: estimate $\langle L \rangle$ by averaging over a much smaller number of samples. This is the basis of the *stochastic gradient descent* (SGD) algorithm. For every step, we take a small set of samples (known as a *batch*) from the training set and compute the gradient of the loss function, averaged over only the samples in the batch. We can view this as an estimate of what we would have gotten if we had averaged over the entire training set, although it may be a very noisy estimate. We perform a single step of gradient descent, then select a new batch of samples for the next step.

This algorithm tends to be much faster. The time required for each step depends only on the size of each batch, which can be quite small (often on the order of 100 samples) and is independent of the size of the training set. The disadvantage is that each step does a less good job of reducing the loss, because it is based on a noisy estimate of the gradient rather than the true gradient. Still, it leads to a much shorter training time overall.

Most optimization algorithms used in deep learning are based on SGD, but there are many variations that improve on it in different ways. Fortunately, you can usually treat these algorithms as black boxes and trust them to do the right thing without understanding all the details of how they work. Two of the most popular algorithms used today are called Adam and RMSProp. If you are in doubt about what algorithm to use, either one of those will probably be a reasonable choice.

Validation

Suppose you have done everything described so far. You collected a large set of training data. You selected a model, then ran a training algorithm until the loss became very small. Congratulations, you now have a function that solves your problem!

Right?

Sorry, it's not that simple! All you really know for sure is that the function works well *on the training data*. You might hope it will also work well on other data, but you certainly can't count on it. Now you need to validate the model to see whether it works on data that it hasn't been specifically trained on.

To do this you need a second dataset, called the *test set*. It has exactly the same form as the training set, a collection of $(\mathbf{x}, \mathbf{y})$ pairs, but the two should have no samples in common. You train the model on the training set, then test it on the test set. This brings us to one of the most important principles in machine learning:

- You must not use the test set in any way while designing or training the model.

In fact, it is best if you never even look at the data in the test set. Test set data is only for testing the fully trained model to find out how well it works. If you allow the test set to influence the model in any way, you risk getting a model that works better on the test set than on other data that was not involved in creating the model. It ceases to be a true test set, and becomes just another type of training set.

This is connected to the mathematical concept of *overfitting*. The training data is supposed to be representative of a much larger data distribution, the set of all inputs you might ever want to use the model on. But you can't train it on all possible inputs. You can only create a finite set of training samples, train the model on those, and hope it learns general strategies that work equally well on other samples. Overfitting is what happens when the training picks up on specific features of the training samples, such that the model works better on them than it does on other samples.

Regularization

Overfitting is a major problem for anyone who uses machine learning. Given that, you won't be surprised to learn that lots of techniques have been developed for avoiding it. These techniques are collectively known as *regularization*. The goal of any regularization technique is to avoid overfitting and produce a trained model that works well on any input, not just the particular inputs that were used for training.

Before we discuss particular regularization techniques, there are two very important points to understand about it.

First, the best way to avoid overfitting is *almost always* to get more training data. The bigger your training set, the better it represents the "true" data distribution, and the less likely the learning algorithm is to overfit. Of course, that is sometimes impossible: maybe you simply have no way to get more data, or the data may be very expensive to collect. In that case, you just have to do the best you can with the data you have, and if overfitting is a problem, you will have to use regularization to avoid it. But more data will probably lead to a better result than regularization.

Second, there is no universally "best" way to do regularization. It all depends on the problem. After all, the training algorithm doesn't know that it's overfitting. All it knows about is the training data. It doesn't know how the true data distribution differs from the training data, so the best it can do is produce a model that works well on the training set. If that isn't what you want, it's up to you to tell it.

That is the essence of any regularization method: biasing the training process to prefer certain types of models over others. You make assumptions about what properties a "good" model should have, and how it differs from an overfit one, and then you tell the training algorithm to prefer models with those properties. Of course, those assumptions are often implicit rather than explicit. It may not be obvious what assumptions you are making by choosing a particular regularization method. But they are always there.

One of the simplest regularization methods is just to train the model for fewer steps. Early in training, it tends to pick up on coarse properties of the training data that likely apply to the true distribution. The longer it runs, the more likely it is to start picking up on fine details of particular training samples. By limiting the number of training steps, you give it less opportunity to overfit. More formally, you are really assuming that "good" parameter values should not be too different from whatever values you start training from.

Another method is to restrict the magnitude of the parameters in the model. For example, you might add a term to the loss function that is proportional to $|\theta|^2$, where θ is a vector containing all of the model's parameters. By doing this, you are assuming that "good" parameter values should not be any larger than necessary. It reflects the fact that overfitting often (though not always) involves some parameters becoming very large.

A very popular method of regularization is called *dropout*. It involves doing something that at first seems ridiculous, but actually works surprisingly well. For each hidden layer in the model, you randomly select a subset of elements in the output vector h_i and set them to 0. On every step of gradient descent, you pick a different random subset of elements. This might seem like it would just break the model: how can you expect it to work when internal calculations keep randomly getting set to 0? The mathematical theory for why dropout works is a bit complicated. Very roughly speaking, by using dropout you are assuming that no individual calculation within the model should be too important. You should be able to randomly remove any individual calculation, and the rest of the model should continue to work without it. This forces it to learn redundant, highly distributed representations of data that make overfitting unlikely. If you are unsure of what regularization method to use, dropout is a good first thing to try.

Hyperparameter Optimization

By now you have probably noticed that there are a lot of choices to make, even when using a supposedly generic model with a "generic" learning algorithm. Examples include:

- The number of layers in the model
- The width of each layer
- The number of training steps to perform
- The learning rate to use during training
- The fraction of elements to set to 0 when using dropout

These options are called *hyperparameters*. A hyperparameter is any aspect of the model or training algorithm that must be set in advance rather than being learned by the training algorithm. But how are you supposed to choose them—and isn't the whole point of machine learning to select settings automatically based on data?

This brings us to the subject of *hyperparameter optimization*. The simplest way of doing it is just to try lots of values for each hyperparameter and see what works best. This becomes very expensive when you want to try lots of values for lots of hyperparameters, so there are more sophisticated approaches, but the basic idea remains the same: try different combinations and see what works best.

But how can you tell what works best? The simplest answer would be to just see what produces the lowest value of the loss function (or some other measure of accuracy) on the training set. But remember, that isn't what we really care about. We want to minimize error on the test set, not the training set. This is especially important for hyperparameters that affect regularization, such as the dropout rate. A low training set error might just mean the model is overfitting, optimizing for the precise details of the training data. So instead we want to try lots of hyperparameter values, then use the ones that minimize the loss on the test set.

But we mustn't do that! Remember: you must not use the test set in any way while designing or training the model. Its job is to tell you how well the model is likely to work on new data it has never seen before. Just because a particular set of hyperparameters happens to work best on the test set doesn't guarantee those values will always work best. We must not allow the test set to influence the model, or it is no longer an unbiased test set.

The solution is to create yet another dataset, which is called the *validation set*. It must not share any samples with either the training set or the test set. The full procedure now works as follows:

1. For each set of hyperparameter values, train the model on the training set, then compute the loss on the validation set.

2. Whichever set of hyperparameters give the lowest loss on the validation set, accept them as your final model.

3. Evaluate that final model on the test set to get an unbiased measure of how well it works.

Other Types of Models

This still leaves one more decision you need to make, and it is a huge subject in itself: what kind of model to use. Earlier in this chapter we introduced multilayer perceptrons. They have the advantage of being a generic class of models that can be applied to many different problems. Unfortunately, they also have serious disadvantages. They require a huge number of parameters, which makes them very susceptible to overfitting. They become difficult to train when they have more than one or two hidden layers. In many cases, you can get a better result by using a less generic model that takes advantage of specific features of your problem.

Much of the content of this book consists of discussing particular types of models that are especially useful in the life sciences. Those can wait until later chapters. But for the purposes of this introduction, there are two very important classes of models we should discuss that are widely used in many different fields. They are called convolutional neural networks and recurrent neural networks.

Convolutional Neural Networks

Convolutional neural networks (CNNs for short) were one of the very first classes of deep models to be widely used. They were developed for use in image processing and computer vision. They remain an excellent choice for many kinds of problems that involve continuous data sampled on a rectangular grid: audio signals (1D), images (2D), volumetric MRI data (3D), and so on.

They are also a class of models that truly justify the term "neural network." The design of CNNs was originally inspired by the workings of the feline visual cortex. (Cats have played a central role in deep learning from the dawn of the field.) Research performed from the 1950s to the 1980s revealed that vision is processed through a series of layers. Each neuron in the first layer takes input from a small region of the visual field (its *receptive field*). Different neurons are specialized to detect particular local patterns or features, such as vertical or horizontal lines. Cells in the second layer take input from local clusters of cells in the first layer, combining their signals to detect more complicated patterns over a larger receptive field. Each layer can be viewed as a new representation of the original image, described in terms of larger and more abstract patterns than the ones in the previous layer.

CNNs mirror this design, sending an input image through a series of layers. In that sense, they are just like MLPs, but the structure of each layer is very different. MLPs use *fully connected layers*. Every element of the output vector depends on every element of the input vector. CNNs use *convolutional layers* that take advantage of spatial locality. Each output element corresponds to a small region of the image, and only depends on the input values in that region. This enormously reduces the number of parameters defining each layer. In effect, it assumes that most elements of the weight matrix $\mathbf{M}_i$ are 0, since each output element only depends on a small number of input elements.

Convolutional layers take this a step further: they assume the parameters are the same for *every local region of the image*. If a layer uses one set of parameters to detect horizontal lines at one location in the image, it also uses exactly the same parameters to detect horizontal lines everywhere else in the image. This makes the number of parameters for the layer independent of the size of the image. All it has to learn is a single *convolutional kernel* that defines how output features are computed from any local region of the image. That local region is often very small, perhaps 5 by 5 pixels. In that case, the number of parameters to learn is only 25 times the number of output features for each region. This is tiny compared to the number in a fully connected layer, making CNNs much easier to train and much less susceptible to overfitting than MLPs.

Recurrent Neural Networks

Recurrent neural networks (RNNs for short) are a bit different. They are normally used to process data that takes the form of a sequence of elements: words in a text document, bases in a DNA molecule, etc. The elements in the sequence are fed into the network's input one at a time. But then the network does something very different: the output from each layer is fed back into its own input on the next step! This allows RNNs to have a sort of memory. When an element (word, DNA base, etc.) from the sequence is fed into the network, the input to each layer depends on that element, but also on all of the previous elements (Figure 2-4).

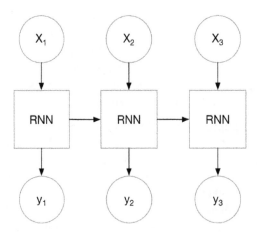

Figure 2-4. A recurrent neural network. As each element $(x_1, x_2, ...)$ of the sequence is fed into the input, the output $(y_1, y_2, ...)$ depends both on the input element and on the RNN's own output during the previous step.

So, the input to a recurrent layer has two parts: the regular input (that is, the output from the previous layer in the network) and the recurrent input (which equals its own output from the previous step). It then needs to calculate a new output based on those inputs. In principle you could use a fully connected layer, but in practice that usually doesn't work very well. Researchers have developed other types of layers that work much better in RNNs. The two most popular ones are called the *gated recurrent unit* (GRU) and the *long short-term memory* (LSTM). Don't worry about the details for now; just remember that if you are creating an RNN, you should usually build it out of one of those types of layers.

Having memory makes RNNs fundamentally different from the other models we have discussed. With a CNN or MLP, you simply feed a value into the network's input and get a different value out. The output is entirely determined by the input. Not so with an RNN. The model has its own internal state, composed of the outputs of all its layers from the most recent step. Each time you feed a new value into the model, the output depends not just on the input value but also on the internal state. Likewise, the internal state is altered by each new input value. This makes RNNs very powerful, and allows them to be used for lots of different applications.

Further Reading

Deep learning is a huge subject, and this chapter has only given the briefest introduction to it. It should be enough to help you read and understand the rest of this book, but if you plan to do serious work in the field, you will want to acquire a much more thorough background. Fortunately, there are many excellent deep learning resources available online. Here are some suggestions for material you might consult:

- *Neural Networks and Deep Learning (http://neuralnetworksanddeeplearning.com)* by Michael Nielsen (Determination Press) covers roughly the same material as this chapter, but goes into far more detail on every subject. If you want a solid working knowledge of the fundamentals of deep learning, sufficient to make use of it in your own work, this is an excellent place to start.

- *Deep Learning (http://www.deeplearningbook.org)* by Ian Goodfellow, Yoshua Bengio, and Aaron Courville (MIT Press) is a more advanced introduction written by some of the top researchers in the field. It expects the reader to have a background similar to that of a graduate student in computer science and goes into far more detail on the mathematical theory behind the subject. You can easily use deep models without understanding all of the theory, but if you want to do original research in deep learning (rather than just using deep models as a tool to solve problems in other fields), this book is a fantastic resource.

- *TensorFlow for Deep Learning* by Bharath Ramsundar and Reza Zadeh (O'Reilly) provides a practitioner's introduction to deep learning that seeks to build intuition about the core concepts without delving too deeply into the mathematical underpinnings of such models. It might be a useful reference for those who are interested in the practical aspects of deep learning.

## CHAPTER 3

Machine Learning with DeepChem

This chapter provides a brief introduction to machine learning with DeepChem, a library built on top of libraries like TensorFlow and PyTorch to facilitate the use of deep learning in the life sciences. DeepChem provides a large collection of models, algorithms, and datasets that are suited to applications in the life sciences. In the remainder of this book, we will use DeepChem to perform our case studies.

Why Not Just Use Keras, TensorFlow, or PyTorch?

This is a common question. The short answer is that the developers of these packages focus their attention on supporting certain types of use cases that prove useful to their core users. For example, there's extensive support for image processing, text handling, and speech analysis. But there's often not a similar level of support in these libraries for molecule handling, genetic datasets, or microscopy datasets. The goal of DeepChem is to give these applications first-class support in the library. This means adding custom deep learning primitives, support for needed file types, and extensive tutorials and documentation for these use cases.

DeepChem is also designed to be well integrated with the TensorFlow and PyTorch ecosystems, so you should be able to mix and match DeepChem code with your other application code based on these frameworks.

In the rest of this chapter, we will assume that you have DeepChem installed on your machine and that you are ready to run the examples. If you don't have DeepChem installed, never fear. Just head over to the DeepChem website (*https://deepchem.io/*) and follow the installation directions for your system.

DeepChem Datasets

DeepChem uses the basic abstraction of the`Dataset` object to wrap the data it uses for machine learning. A `Dataset` contains the information about a set of samples: the input vectors x, the target output vectors y, and possibly other information such as a description of what each sample represents. There are subclasses of `Dataset` corresponding to different ways of storing the data. The `NumpyDataset` object in particular serves as a convenient wrapper for NumPy arrays and will be used extensively. In this section, we will walk through a simple code case study of how to use `NumpyDataset`. All of this code can be entered in the interactive Python interpreter; where appropriate, the output is shown.

We start with some simple imports:

```
import deepchem as dc
import numpy as np
```

Let's now construct some simple NumPy arrays:

```
x = np.random.random((4, 5))
y = np.random.random((4, 1))
```

This dataset will have four samples. The array x has five elements ("features") for each sample, and y has one element for each sample. Let's take a quick look at the actual arrays we've sampled (note that when you run this code locally, you should expect to see different numbers since your random seed will be different):

```
In : x
Out:
array([[0.960767 , 0.31300931, 0.23342295, 0.59850938, 0.30457302],
    [0.48891533, 0.69610528, 0.02846666, 0.20008034, 0.94781389],
    [0.17353084, 0.95867152, 0.73392433, 0.47493093, 0.4970179 ],
    [0.15392434, 0.95759308, 0.72501478, 0.38191593, 0.16335888]])

In : y
Out:
array([[0.00631553],
    [0.69677301],
    [0.16545319],
    [0.04906014]])
```

Let's now wrap these arrays in a `NumpyDataset` object:

```
dataset = dc.data.NumpyDataset(x, y)
```

We can unwrap the `dataset` object to get at the original arrays that we stored inside:

```
In : print(dataset.X)
[[0.960767 0.31300931 0.23342295 0.59850938 0.30457302]
 [0.48891533 0.69610528 0.02846666 0.20008034 0.94781389]
 [0.17353084 0.95867152 0.73392433 0.47493093 0.4970179 ]
```

```
[0.15392434 0.95759308 0.72501478 0.38191593 0.16335888]]

In : print(dataset.y)
[[0.00631553]
[0.69677301]
[0.16545319]
[0.04906014]]
```

Note that these arrays are the same as the original arrays x and y:

```
In : np.array_equal(x, dataset.X)
Out : True

In : np.array_equal(y, dataset.y)
Out : True
```

Other Types of Datasets

DeepChem has support for other types of Dataset objects, as mentioned previously. These types primarily become useful when dealing with larger datasets that can't be entirely stored in computer memory. There is also integration for DeepChem to use TensorFlow's tf.data dataset loading utilities. We will touch on these more advanced library features as we need them.

Training a Model to Predict Toxicity of Molecules

In this section, we will demonstrate how to use DeepChem to train a model to predict the toxicity of molecules. In a later chapter, we will explain how toxicity prediction for molecules works in much greater depth, but in this section, we will treat it as a black-box example of how DeepChem models can be used to solve machine learning challenges. Let's start with a pair of needed imports:

```
import numpy as np
import deepchem as dc
```

The next step is loading the associated toxicity datasets for training a machine learning model. DeepChem maintains a module called dc.molnet (short for MoleculeNet) that contains a number of preprocessed datasets for use in machine learning experimentation. In particular, we will make use of the dc.molnet.load_tox21() function, which will load and process the Tox21 toxicity dataset for us. When you run these commands for the first time, DeepChem will process the dataset locally on your machine. You should expect to see processing notes like the following:

```
In : tox21_tasks, tox21_datasets, transformers = dc.molnet.load_tox21()
Out: Loading raw samples now.
shard_size: 8192
About to start loading CSV from /tmp/tox21.CSV.gz
Loading shard 1 of size 8192.
```

```
Featurizing sample 0
Featurizing sample 1000
Featurizing sample 2000
Featurizing sample 3000
Featurizing sample 4000
Featurizing sample 5000
Featurizing sample 6000
Featurizing sample 7000
TIMING: featurizing shard 0 took 15.671 s
TIMING: dataset construction took 16.277 s
Loading dataset from disk.
TIMING: dataset construction took 1.344 s
Loading dataset from disk.
TIMING: dataset construction took 1.165 s
Loading dataset from disk.
TIMING: dataset construction took 0.779 s
Loading dataset from disk.
TIMING: dataset construction took 0.726 s
Loading dataset from disk.
```

The process of *featurization* is how a dataset containing information about molecules is transformed into matrices and vectors for use in machine learning analyses. We will explore this process in greater depth in subsequent chapters. Let's start here, though, by taking a quick peek at the data we've processed.

The `dc.molnet.load_tox21()` function returns multiple outputs: `tox21_tasks`, `tox21_datasets`, and `transformers`. Let's briefly take a look at each:

```
In : tox21_tasks
Out:
['NR-AR',
 'NR-AR-LBD',
 'NR-AhR',
 'NR-Aromatase',
 'NR-ER',
 'NR-ER-LBD',
 'NR-PPAR-gamma',
 'SR-ARE',
 'SR-ATAD5',
 'SR-HSE',
 'SR-MMP',
 'SR-p53']

In : len(tox21_tasks)
Out: 12
```

Each of the 12 tasks here corresponds with a particular biological experiment. In this case, each of these tasks is for an *enzymatic assay* which measures whether the molecules in the Tox21 dataset bind with the *biological target* in question. The terms NR-AR and so on correspond with these targets. In this case, each of these targets is a particular enzyme believed to be linked to toxic responses to potential therapeutic molecules.

How Much Biology Do I Need to Know?

For computer scientists and engineers entering the life sciences, the array of biological terms can be dizzying. However, it's not necessary to have a deep understanding of biology in order to begin making an impact in the life sciences. If your primary background is in computer science, it can be useful to try understanding biological systems in terms of computer scientific analogues. Imagine that cells or animals are complex legacy codebases that you have no control over. As an engineer, you have a few experimental measurements of these systems (assays) which you can use to gain some understanding of the underlying mechanics. Machine learning is an extraordinarily powerful tool for understanding biological systems since learning algorithms are capable of extracting useful correlations in a mostly automatic fashion. This allows even biological beginners to sometimes find deep biological insights.

In the remainder of this book, we discuss basic biology in brief asides. These notes can serve as entry points into the vast biological literature. Public references such as Wikipedia often contain a wealth of useful information, and can help bootstrap your biological education.

Next, let's consider `tox21_datasets`. The use of the plural is a clue that this field is actually a tuple containing multiple`dc.data.Dataset` objects:

```
In : tox21_datasets
Out:
(<deepchem.data.datasets.DiskDataset at 0x7f9804d6c390>,
 <deepchem.data.datasets.DiskDataset at 0x7f9804d6c780>,
 <deepchem.data.datasets.DiskDataset at 0x7f9804c5a518>)
```

In this case, these datasets correspond to the training, validation, and test sets you learned about in the previous chapter. You might note that these are `DiskDataset` objects; the `dc.molnet` module caches these datasets on your disk so that you don't need to repeatedly refeaturize the Tox21 dataset. Let's split up these datasets correctly:

```
train_dataset, valid_dataset, test_dataset = tox21_datasets
```

When dealing with new datasets, it's very useful to start by taking a look at their shapes. To do so, inspect the `shape` attribute:

```
In : train_dataset.X.shape
Out: (6264, 1024)

In : valid_dataset.X.shape
```

```
Out: (783, 1024)

In : test_dataset.X.shape
Out: (784, 1024)
```

The train_dataset contains a total of 6,264 samples, each of which has an associated feature vector of length 1,024. Similarly, valid_dataset and test_datasetcontain respectively 783 and 784 samples. Let's now take a quick look at the y vectors for these datasets:

```
In : np.shape(train_dataset.y)
Out: (6264, 12)

In : np.shape(valid_dataset.y)
Out: (783, 12)

In : np.shape(test_dataset.y)
Out: (784, 12)
```

There are 12 data points, also known as *labels*, for each sample. These correspond to the 12 tasks we discussed earlier. In this particular dataset, the samples correspond to molecules, the tasks correspond to biochemical assays, and each label is the result of a particular assay on a particular molecule. Those are what we want to train our model to predict.

There's a complication, however: the actual experimental dataset for Tox21 did not test every molecule in every biological experiment. That means that some of these labels are meaningless placeholders. We simply don't have any data for some properties of some molecules, so we need to ignore those elements of the arrays when training and testing the model.

How can we find which labels were actually measured? We can check the dataset's w field, which records its *weights*. Whenever we compute the loss function for a model, we multiply by w before summing over tasks and samples. This can be used for a few purposes, one being to flag missing data. If a label has a weight of 0, that label does not affect the loss and is ignored during training. Let's do some digging to find how many labels have actually been measured in our datasets:

```
In : train_dataset.w.shape
Out: (6264, 12)

In : np.count_nonzero(train_dataset.w)
Out: 62166

In : np.count_nonzero(train_dataset.w == 0)
Out: 13002
```

Of the $6,264 \times 12 = 75,168$ elements in the array of labels, only 62,166 were actually measured. The other 13,002 correspond to missing measurements and should be ignored. You might ask, then, why we still keep such entries around. The answer is mainly for convenience; irregularly shaped arrays are much harder to reason about and deal with in code than regular matrices with an associated set of weights.

Processing Datasets Is Challenging

It's important to note here that cleaning and processing a dataset for use in the life sciences can be extremely challenging. Many raw datasets will contain systematic classes of errors. If the dataset in question has been constructed from an experiment conducted by an external organization (a contract research organization, or CRO), it's quite possible that the dataset will be systematically wrong. For this reason, many life science organizations maintain scientists in-house whose job it is to verify and clean such datasets.

In general, if your machine learning algorithm isn't working for a life science task, there's a significant chance that the root cause stems not from the algorithm but from systematic errors in the source of data that you're using.

Now let's examine `transformers`, the final output that was returned by `load_tox21()`. A *transformer* is an object that modifies a dataset in some way. Deep-Chem provides many transformers that manipulate data in useful ways. The data-loading routines found in MoleculeNet always return a list of transformers that have been applied to the data, since you may need them later to "untransform" the data. Let's see what we have in this case:

```
In : transformers
Out: [<deepchem.trans.transformers.BalancingTransformer at 0x7f99dd73c6d8>]
```

Here, the data has been transformed with a `BalancingTransformer`. This class is used to correct for unbalanced data. In the case of Tox21, most molecules do not bind to most of the targets. In fact, over 90% of the labels are 0. That means a model could trivially achieve over 90% accuracy simply by always predicting 0, no matter what input it was given. Unfortunately, that model would be completely useless! Unbalanced data, where there are many more training samples for some classes than others, is a common problem in classification tasks.

Fortunately, there is an easy solution: adjust the dataset's matrix of weights to compensate. `BalancingTransformer` adjusts the weights for individual data points so that the total weight assigned to every class is the same. That way, the loss function has no systematic preference for any one class. The loss can only be decreased by learning to correctly distinguish between classes.

Now that we've explored the Tox21 datasets, let's start exploring how we can train models on these datasets. DeepChem's `dc.models` submodule contains a variety of different life science–specific models. All of these various models inherit from the parent class `dc.models.Model`. This parent class is designed to provide a common API that follows common Python conventions. If you've used other Python machine learning packages, you should find that many of the `dc.models.Model` methods look quite familiar.

In this chapter, we won't really dig into the details of how these models are constructed. Rather, we will just provide an example of how to instantiate a standard Deep-Chem model, `dc.models.MultitaskClassifier`. This model builds a fully connected network (an MLP) that maps input features to multiple output predictions. This makes it useful for *multitask* problems, where there are multiple labels for every sample. It's well suited for our Tox21 datasets, since we have a total of 12 different assays we wish to predict simultaneously. Let's see how we can construct a `MultitaskClassifier` in DeepChem:

```
model = dc.models.MultitaskClassifier(n_tasks=12,
    n_features=1024,
    layer_sizes=[1000])
```

There are a variety of different options here. Let's briefly review them. `n_tasks` is the number of tasks, and `n_features` is the number of input features for each sample. As we saw earlier, the Tox21 dataset has 12 tasks and 1,024 features for each sample. `layer_sizes` is a list that sets the number of fully connected hidden layers in the network, and the width of each one. In this case, we specify that there is a single hidden layer of width 1,000.

Now that we've constructed the model, how can we train it on the Tox21 datasets? Each `Model` object has a `fit()` method that fits the model to the data contained in a `Dataset` object. Fitting our `MultitaskClassifier` object is then a simple call:

```
model.fit(train_dataset, nb_epoch=10)
```

Note that we added on a flag here. `nb_epoch=10` says that 10 epochs of gradient descent training will be conducted. An *epoch* refers to one complete pass through all the samples in a dataset. To train a model, you divide the training set into batches and take one step of gradient descent for each batch. In an ideal world, you would reach a well-optimized model before running out of data. In practice, there usually isn't enough training data for that, so you run out of data before the model is fully trained. You then need to start reusing data, making additional passes through the dataset. This lets you train models with smaller amounts of data, but the more epochs you use, the more likely you are to end up with an overfit model.

Let's now evaluate the performance of the trained model. In order to evaluate how well a model works, it is necessary to specify a metric. The DeepChem class `dc.metrics.Metric` provides a general way to specify metrics for models. For the Tox21 datasets, the ROC AUC score is a useful metric, so let's do our analysis using it. However, note a subtlety here: there are multiple Tox21 tasks. Which one do we compute the ROC AUC on? A good tactic is to compute the mean ROC AUC score across all tasks. Luckily, it's easy to do this:

```
metric = dc.metrics.Metric(dc.metrics.roc_auc_score, np.mean)
```

Since we've specified `np.mean`, the mean of the ROC AUC scores across all tasks will be reported. DeepChem models support the evaluation function `model.evaluate()`, which evaluates the performance of the model on a given dataset and metric:

 ROC AUC

We want to classify molecules as toxic or nontoxic, but the model outputs continuous numbers, not discrete predictions. In practice, you pick a threshold value and predict that a molecule is toxic whenever the output is greater than the threshold. A low threshold will produce many false positives (predicting a safe molecule is actually toxic). A higher threshold will give fewer false positives but more false negatives (incorrectly predicting that a toxic molecule is safe).

The *receiver operating characteristic* (ROC) curve is a convenient way to visualize this trade-off. You try many different threshold values, then plot a curve of the true positive rate versus the false positive rate as the threshold is varied. An example is shown in Figure 3-1.

The ROC AUC is the total area under the ROC curve. The *area under the curve* (AUC) provides an indication of the model's ability to distinguish different classes. If there exists any threshold value for which every sample is classified correctly, the ROC AUC score is 1. At the other extreme, if the model outputs completely random values unrelated to the true classes, the ROC AUC score is 0.5. This makes it a useful number for summarizing how well a classifier works. It's just a heuristic, but it's a popular one.

```
train_scores = model.evaluate(train_dataset, [metric], transformers)
test_scores = model.evaluate(test_dataset, [metric], transformers)
```

Now that we've calculated the scores, let's take a look!

```
In : print(train_scores)
...: print(test_scores)
Out
{'mean-roc_auc_score': 0.9582370607146093}
{'mean-roc_auc_score': 0.6826331496516941}
```

Notice that our score on the training set (0.96) is much better than our score on the test set (0.68). This shows the model has been overfit. The test set score is the one we really care about. These numbers aren't the best possible on this dataset, but they aren't bad at all for an out-of-the-box system. The complete ROC curve for one of the 12 tasks is shown in Figure 3-1.

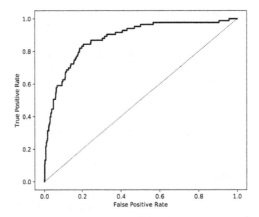

Figure 3-1. The ROC curve for one of the 12 tasks. The dotted diagonal line shows what the curve would be for a model that just guessed at random. The actual curve is consistently well above the diagonal, showing that we are doing much better than random guessing.

Case Study: Training an MNIST Model

In the previous section, we covered the basics of training a machine learning model with DeepChem. However, we used a premade model class, `dc.models.Multitask Classifier`. Sometimes you may want to create a new deep learning architecture instead of using a preconfigured one. In this section, we discuss how to train a convolutional neural network on the MNIST digit recognition dataset. Instead of using a premade architecture like in the previous example, this time we will specify the full deep learning architecture ourselves. To do so, we will use the Keras modeling API that is built into TensorFlow.

When Do Canned Models Make Sense?

In this section, we're going to use a custom architecture on MNIST. In the previous example, we used a "canned" (that is, predefined) architecture instead. When does each alternative make sense? If you have a well-debugged canned architecture for a problem, it will likely make sense to use it. But if you're working on a new dataset where no such architecture has been put together, you'll often have to create a custom architecture. It's important to be familiar with using both canned and custom architectures, so we've included an example of each in this chapter.

The MNIST Digit Recognition Dataset

The MNIST digit recognition dataset (see Figure 3-2) requires the construction of a machine learning model that can learn to classify handwritten digits correctly. The challenge is to classify digits from 0 to 9 given 28 × 28-pixel black and white images. The dataset contains 60,000 training examples and a test set of 10,000 examples.

Figure 3-2. Samples drawn from the MNIST handwritten digit recognition dataset. (Source: GitHub (https://github.com/mnielsen/rmnist/blob/master/data/ rmnist_10.png))

The MNIST dataset is not particularly challenging as far as machine learning problems go. Decades of research have produced state-of-the-art algorithms that achieve close to 100% test set accuracy on this dataset. As a result, the MNIST dataset is no longer suitable for research work, but it is a good tool for pedagogical purposes.

Isn't DeepChem Just for the Life Sciences?

As we mentioned earlier in the chapter, it's entirely feasible to use other deep learning packages for life science applications. Similarly, it's possible to build general machine learning systems using Deep-Chem. Although building a movie recommendation system in DeepChem might be trickier than it would be with more specialized tools, it would be quite feasible to do so. And for good reason: there have been multiple studies looking into the use of recommendation system algorithms for use in molecular binding prediction. Machine learning architectures used in one field tend to carry over to other fields, so it's important to retain the flexibility needed for innovative work.

A Convolutional Architecture for MNIST

DeepChem lets you use either Keras/TensorFlow or PyTorch to construct nonstandard deep learning architectures. In this section, we will walk through the code required to construct the convolutional architecture shown in Figure 3-3 using Keras. It begins with two convolutional layers to identify local features within the image. They are followed by two fully connected layers to predict the digit from those local features.

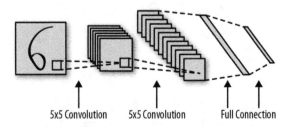

5x5 Convolution   5x5 Convolution   Full Connection

Figure 3-3. An illustration of the architecture that we will construct in this section for processing the MNIST dataset.

Let's start with the necessary imports:

```
import deepchem as dc
import tensorflow as tf
import tensorflow.keras.layers as layers
```

The submodule `tensorflow.keras.layers` contains a collection of "layers." These layers serve as building blocks of deep architectures and can be composed to build new deep learning architectures. We will demonstrate how layer objects are used shortly.

First we need to load the MNIST dataset. Since it is such a popular example, Keras provides a built in copy.

```
(x_train, y_train), (x_test, y_test) = tf.keras.datasets.mnist.load_data()
y_train = tf.one_hot(y_train, 10).numpy()
y_test = tf.one_hot(y_test, 10).numpy()
```

Notice how we call `tf.one_hot()` to convert the labels to one-hot encoding. Keras provides the labels as a single number for each sample, which is not the representation we want to use as the input to our model.

One-Hot Encoding

The MNIST dataset is categorical. That is, objects belong to one of a finite list of potential categories. In this case, these categories are the digits 0 through 9. How can we feed these categories into a machine learning system? One obvious answer would be to simply feed in a single number that takes values from 0 through 9. However, for a variety of technical reasons, this encoding often doesn't seem to work well. The alternative that people commonly use is to *one-hot encode*. Each label for MNIST is a vector of length 10 in which a single element is set to 1, and all others are set to 0. If the nonzero value is at the 0th index, then the label corresponds to the digit 0. If the nonzero value is at the 9th index, then the label corresponds to the digit 9.

Next, we construct `NumpyDataset` objects that wrap the MNIST training and test datasets:

```
train_dataset = dc.data.NumpyDataset(x_train, y_train)
test_dataset = dc.data.NumpyDataset(x_test, y_test)
```

With the training and test datasets in hand, we can now turn our attention towards defining the architecture for the MNIST convolutional network.

The key concept this is based on is that layer objects can be composed to build new models. As we discussed in the previous chapter, each layer takes input from previous layers and computes an output that can be passed to subsequent layers. At the very start, there are input layers that take in features. At the other end are output layers that return the results of the performed computation. In this example, we will compose a sequence of layers in order to construct an image-processing convolutional network.

Let's start by adding some inputs for features by using the `Input` class:

```
features = tf.keras.Input(shape=(28, 28, 1))
```

MNIST contains images of size 28 × 28. The third dimension is the number of features for each pixel. MNIST has black and white images, so there is only a single feature. If they were color images, the third dimension would be 3 corresponding to the three color channels.

Now we can pass it through to the convolutional layers:

```
conv2d_1 = layers.Conv2D(filters=32, kernel_size=5,
                         activation=tf.nn.relu)(features)
conv2d_2 = layers.Conv2D(filters=64, kernel_size=5,
                         activation=tf.nn.relu)(conv2d_1)
```

Here, the `Conv2D` class applies a 2D convolution to each sample of its input, then passes it through a rectified linear unit (ReLU) activation function. Note how we construct each layer then invoke it, passing the output of the previous layer as an argument. We want to end by applying `Dense` (fully connected) layers to the outputs of the convolutional layer. However, the output of `Conv2D` layers is 2D, so we will first need to apply a `Flatten` layer to flatten our input to one dimension:

```
flatten = layers.Flatten()(conv2d_2)
dense1 = layers.Dense(units=1024, activation=tf.nn.relu)(flatten)
dense2 = layers.Dense(units=10, activation=None)(dense1)
```

The `units` argument in a `Dense` layer specifies the width of the layer. The first layer outputs 1,024 values per sample, but the second layer outputs 10 values, corresponding to our 10 possible digit values.

We are almost ready to wrap this up in a `tf.keras.Model` object. First let's consider the outputs and loss function for our model.

SoftMax and SoftMaxCrossEntropy

You often want a model to output a probability distribution. For MNIST, we want to output the probability that a given sample represents each of the 10 digits. Every output must be positive, and they must sum to 1. An easy way to achieve this is to let the model compute arbitrary numbers, then pass them through the confusingly named *softmax* function:

$$\sigma_i(x) = \frac{e^{x_i}}{\Sigma_j e^{x_j}}$$

The exponential in the numerator ensures that all values are positive, and the sum in the denominator ensures they add up to 1. If one element of x is much larger than the others, the corresponding output element is very close to 1 and all the other outputs are very close to 0.

Let's use the softmax function to compute output probabilities:

```
output = layers.Activation(tf.math.softmax)(dense2)
```

The `SoftmaxCrossEntropy` loss function first uses a softmax function to convert the outputs to probabilities, then computes the cross entropy of those probabilities with the labels. Remember that the labels are one-hot encoded: 1 for the correct class, 0 for all others. You can think of that as a probability distribution! The loss is minimized when the predicted probability of the correct class is as close to 1 as possible. These two operations (softmax followed by cross entropy) often appear together, and computing them as a single step turns out to be more numerically stable than performing them separately.

We want the model to return `output` when we use it to make predictions, but for numerical stability we want it to use `dense2` as the input to the loss function. So let's have it return both of them.

```
keras_model = tf.keras.Model(inputs=features, outputs=[output, dense2])
```

The final step is to wrap the `tf.keras.Model` in a `dc.models.KerasModel`.

```
model = dc.models.KerasModel(
    keras_model,
    loss=dc.models.losses.SoftmaxCrossEntropy(),
    output_types=['prediction', 'loss'],
    model_dir='mnist')
```

The `output_types` argument tells it how to interpret each of the outputs produced by the model. The first one has type "prediction", so it will be returned when you call `predict()`, and the second one has type "loss", meaning that it should be passed to the loss function. For simpler models where a single output is used for both purposes, this argument can be omitted.

The `model_dir` option specifies a directory where the model's parameters should be saved. You can omit this, as we did in the previous example, but then the model will not be saved. As soon as the Python interpreter exits, all your hard work training the model will be thrown out! Specifying a directory allows you to reload the model later and make new predictions with it.

Note that since `KerasModel` inherits from `Model`, this object is an instance of `dc.models.Model` and supports the same `fit()` and `evaluate()` functions we saw previously:

```
In : isinstance(model, dc.models.Model)
Out: True
```

We can now train the model using the same `fit()` function we called in the previous section:

```
model.fit(train_dataset, nb_epoch=10)
```

Note that this method call might take some time to execute on a standard laptop! If the function is not executing quickly enough, try using `nb_epoch=1`. The results will be worse, but you will be able to complete the rest of this chapter more quickly.

Let's define our metric this time to be accuracy, the fraction of labels that are correctly predicted:

```
metric = dc.metrics.Metric(dc.metrics.accuracy_score)
```

We can then compute the accuracy using the same computation as before:

```
train_scores = model.evaluate(train_dataset, [metric])
test_scores = model.evaluate(test_dataset, [metric])
```

This produces excellent performance: the accuracy is 0.999 on the training set, and 0.991 on the test set. Our model identifies more than 99% of the test set samples correctly.

Try to Get Access to a GPU

As you saw in this chapter, deep learning code can run pretty slowly! Training a convolutional neural network on a good laptop can take more than an hour to complete. This is because this code depends on a large number of linear algebraic operations on image data. Most CPUs are not well equipped to perform these types of computations.

If possible, try to get access to a modern graphics processing unit. These cards were originally developed for gaming, but are now used for many types of numeric computations. Most modern deep learning workloads will run much faster on GPUs. The examples you'll see in this book will be easier to complete with GPUs as well.

If it's not feasible to get access to a GPU, don't worry. You'll still be able to complete the exercises in this book—they might just take a little longer (you might have to grab a coffee or read a book while you wait for the code to finish running).

Conclusion

In this chapter, you've learned how to use the DeepChem library to implement some simple machine learning systems. In the remainder of this book, we will continue to use DeepChem as our library of choice, so don't worry if you don't have a strong grasp of the fundamentals of the library yet. There will be plenty more examples coming.

In subsequent chapters, we will begin to introduce the basic concepts needed to do effective machine learning on life science datasets. In the next chapter, we will introduce you to machine learning on molecules.

## CHAPTER 4

Machine Learning for Molecules

This chapter covers the basics of performing machine learning on molecular data. Before we dive into the chapter, it might help for us to briefly discuss why molecular machine learning can be a fruitful subject of study. Much of modern materials science and chemistry is driven by the need to design new molecules that have desired properties. While significant scientific work has gone into new design strategies, much random search is sometimes still needed to construct interesting molecules. The dream of molecular machine learning is to replace such random experimentation with guided search, where machine-learned predictors can propose which new molecules might have desired properties. Such accurate predictors could enable the creation of radically new materials and chemicals with useful properties.

This dream is compelling, but how can we get started on this path? The first step is to construct technical methods for transforming molecules into vectors of numbers that can then be passed to learning algorithms. Such methods are called *molecular featurizations*. We will cover a number of them in this chapter, and more in the next chapter. Molecules are complex entities, and researchers have developed a host of different techniques for featurizing them. These representations include chemical descriptor vectors, 2D graph representations, 3D electrostatic grid representations, orbital basis function representations, and more.

Once featurized, a molecule still needs to be learned from. We will review some algorithms for learning functions on molecules, including simple fully connected networks as well as more sophisticated techniques like graph convolutions. We'll also describe some of the limitations of graph convolutional techniques, and what we should and should not expect from them. We'll end the chapter with a molecular machine learning case study on an interesting dataset.

What Is a Molecule?

Before we dive into molecular machine learning in depth, it will be useful to review what exactly a molecule is. This question sounds a little silly, since molecules like H_2O and CO_2 are introduced to even young children. Isn't the answer obvious? The fact is, though, that for the vast majority of human existence, we had no idea that molecules existed at all. Consider a thought experiment: how would you convince a skeptical alien that entities called molecules exist? The answer turns out to be quite sophisticated. You might, for example, need to break out a mass spectrometer!

Mass Spectroscopy

Identifying the molecules that are present in a given sample can be quite challenging. The most popular technique at present relies on mass spectroscopy. The basic idea of mass spectroscopy is to bombard a sample with electrons. This bombardment shatters the molecules into fragments. These fragments typically *ionize*—that is, pick up or lose electrons to become charged. These charged fragments are propelled by an electric field which separates them based on their mass-to-charge ratio. The spread of detected charged fragments is called the *spectrum*. Figure 4-1 illustrates this process. From the collection of detected fragments, it is often possible to identify the precise molecules that were in the original sample. However, this process is still lossy and difficult. A number of researchers are actively researching techniques to improve mass spectroscopy with deep learning algorithms to ease the identification of the original molecules from the detected charged spectrum.

Note the complexity of performing this detection! Molecules are complicated entities that are tricky to pin down precisely.

For the sake of getting started, let's presume a definition of a molecule as a group of atoms joined together by physical forces. Molecules are the smallest fundamental unit of a chemical compound that can take part in a chemical reaction. Atoms in a molecule are connected with one another by *chemical bonds*, which hold them together and restrict their motion relative to each other. Molecules come in a huge range of sizes, from just a few atoms up to many thousands of atoms. Figure 4-2 provides a simple depiction of a molecule in this model.

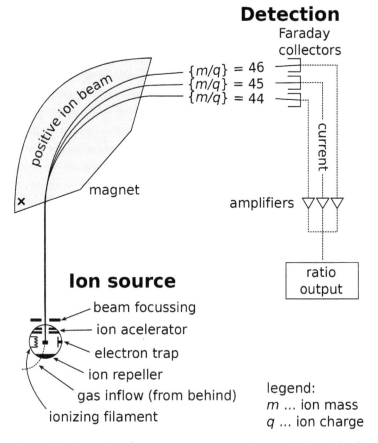

Detection

Faraday
collectors

$\{m/q\} = 46$
$\{m/q\} = 45$
$\{m/q\} = 44$

current

amplifiers ▽▽▽

ratio
output

positive ion beam

✕

magnet

Ion source

- beam focussing
- ion acelerator
- electron trap
- ion repeller
- gas inflow (from behind)
- ionizing filament

legend:
m ... ion mass
q ... ion charge

Figure 4-1. A simple schematic of a mass spectrometer. (Source: Wikimedia (https:// commons.wikimedia.org/wiki/File:Mass_Spectrometer_Schematic.svg).)

Figure 4-2. A simple representation of a caffeine molecule as a "ball-and-stick" diagram. Atoms are represented as colored balls (black is carbon, red is oxygen, blue is nitrogen, white is hydrogen) joined by sticks which represent chemical bonds.

With this basic description in hand, we'll spend the next couple of sections diving into more detail about various aspects of molecular chemistry. It's not critical that you get all of these concepts on your first reading of this chapter, but it can be useful to have some basic knowledge of the chemical landscape at hand.

Molecules Are Dynamic, Quantum Entities

We've just provided a simplistic description of molecules in terms of atoms and bonds. It's very important to keep in the back of your mind that there's a lot more going on within any molecule. For one, molecules are dynamic entities, so all the atoms within a given molecule are in rapid motion with respect to one another. The bonds themselves are stretching back and forth and perhaps oscillating in length rapidly. It's quite common for atoms to rapidly break off from and rejoin molecules. We'll see a bit more about the dynamic nature of molecules shortly, when we discuss molecular conformations.

Even more strangely, molecules are quantum. There are a lot of layers to saying that an entity is quantum, but as a simple description, it's important to note that "atoms" and "bonds" are much less well defined than a simple ball-and-stick diagram might imply. There's a lot of fuzziness in the definitions here. It's not important that you grasp these complexities at this stage, but remember that our depictions of molecules are very approximate. This can have practical relevance, since some learning tasks may require describing molecules with different depictions than others.

What Are Molecular Bonds?

It may have been a while since you studied basic chemistry, so we will spend time reviewing basic chemical concepts here and there. The most basic question is, what is a chemical bond?

The molecules that make up everyday life are made of atoms, often very large numbers of them. These atoms are joined together by chemical bonds. These bonds essentially "glue" together atoms by their shared electrons. There are many different types of molecular bonds, including covalent bonds and several types of noncovalent bonds.

Covalent bonds

Covalent bonds involve sharing electrons between two atoms, such that the same electrons spend time around both atoms (Figure 4-3). In general, covalent bonds are the strongest type of chemical bond. They are formed and broken in chemical reactions. Covalent bonds tend to be very stable: once they form, it takes a lot of energy to

---

break them, so the atoms can remain bonded for a very long time. This is why molecules behave as distinct objects rather than loose collections of unrelated atoms. In fact, covalent bonds are what define molecules: a molecule is a set of atoms joined by covalent bonds.

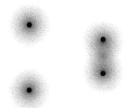

Figure 4-3. Left: two atomic nuclei, each surrounded by a cloud of electrons. Right: as the atoms come close together, the electrons start spending more time in the space between the nuclei. This attracts the nuclei together, forming a covalent bond between the atoms.

Noncovalent bonds

Noncovalent bonds don't involve the direct sharing of electrons between atoms, but they do involve weaker electromagnetic interactions. Since they are not as strong as covalent bonds, they are more ephemeral, constantly breaking and reforming. Noncovalent bonds do not "define" molecules in the same sense that covalent bonds do, but they have a huge effect on determining the shapes molecules take on and the ways different molecules associate with each other.

"Noncovalent bonds" is a generic term covering several different types of interactions. Some examples of noncovalent bonds include hydrogen bonds (Figure 4-4), salt bridges, pi-stacking, and more. These types of interactions often play crucial roles in drug design, since most drugs interact with biological molecules in the human body through noncovalent interactions.

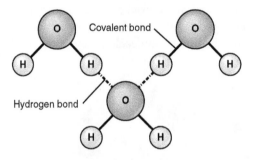

Figure 4-4. Water molecules have strong hydrogen bonding interactions between hydrogen and oxygen on adjacent molecules. A strong network of hydrogen bonds contributes in part to water's power as a solvent. (Source: Wikimedia (https://commons.wikimedia.org/wiki/File:SimpleBayesNet.svg).)

We'll run into each of these types of bonds at various points in the book. In this chapter, we will mostly deal with covalent bonds, but noncovalent interactions will become much more crucial when we start studying some biophysical deep models.

Molecular Graphs

A *graph* is a mathematical data structure made up of *nodes* connected together by *edges* (Figure 4-5). Graphs are incredibly useful abstractions in computer science. In fact, there is a whole branch of mathematics called graph theory dedicated to understanding the properties of graphs and finding ways to manipulate and analyze them. Graphs are used to describe everything from the computers that make up a network, to the pixels that make up an image, to actors who have appeared in movies with Kevin Bacon.

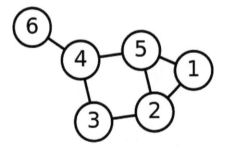

Figure 4-5. An example of a mathematical graph with six nodes connected by edges. (Source: Wikimedia (https://commons.wikimedia.org/wiki/File:6n-graf.svg).)

Importantly, molecules can be viewed as graphs as well (Figure 4-6). In this description, the atoms are the nodes in the graph, and the chemical bonds are the edges. Any molecule can be converted into a corresponding molecular graph.

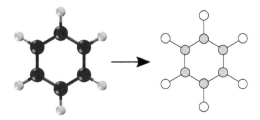

Figure 4-6. An example of converting a benzene molecule into a molecular graph. Note that atoms are converted into nodes and chemical bonds into edges.

In the remainder of this chapter, we will repeatedly convert molecules into graphs in order to analyze them and learn to make predictions about them.

Molecular Conformations

A molecular graph describes the set of atoms in a molecule and how they are bonded together. But there is another very important thing we need to know: how the atoms are positioned relative to each other in 3D space. This is called the molecule's *conformation*.

Atoms, bonds, and conformation are related to each other. If two atoms are covalently bonded, that tends to fix the distance between them, strongly restricting the possible conformations. The angles formed by sets of three or four bonded atoms are also often restricted. Sometimes there will be whole clusters of atoms that are completely rigid, all moving together as a single unit. But other pieces of molecules are flexible, allowing atoms to move relative to each other. For example, many (but not all) covalent bonds allow the groups of atoms they connect to freely rotate around the axis of the bond. This lets the molecule take on many different conformations.

Figure 4-7 shows a very popular molecule: sucrose, also known as table sugar. It is shown both as a 2D chemical structure and as a 3D conformation. Sucrose consists of two rings linked together. Each of the rings is fairly rigid, so its shape changes very little over time. But the linker connecting them is much more flexible, allowing the rings to move relative to each other.

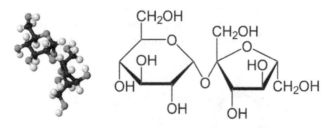

Figure 4-7. Sucrose, represented as a 3D conformation and a 2D chemical structure. (Adapted from Wikimedia images (Wikimedia (https://commons.wikimedia.org/wiki/ File:Sucrose-3D-balls.png) and Wikipedia (https://en.wikipedia.org/wiki/File:Saccha rose2.svg).)

As molecules get larger, the number of feasible conformations they can take grows enormously. For large macromolecules such as proteins (Figure 4-8), computationally exploring the set of possible conformations currently requires very expensive simulations.

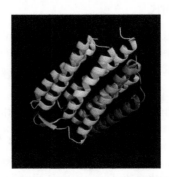

Figure 4-8. A conformation of bacteriorhodopsin (used to capture light energy) rendered in 3D. Protein conformations are particularly complex, with multiple 3D geometric motifs, and serve as a good reminder that molecules have geometry in addition to their chemical formulas. (Source: Wikimedia (https://upload.wikimedia.org/wikipedia/ commons/thumb/d/dd/1M0K.png/480px-1M0K.png).)

Chirality of Molecules

Some molecules (including many drugs) come in two forms that are mirror images of each other. This is called *chirality*. A chiral molecule has both a "right-handed" form (also known as the "R" form) and a "left-handed" form (also known as the "S" form), as illustrated in Figure 4-9.

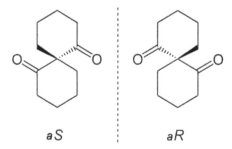

aS   aR

Figure 4-9. Axial chirality of a spiro compound (a compound made up of two or more rings joined together). Note that the two chiral variants are respectively denoted as "R" and "S." This convention is widespread in the chemistry literature.

Chirality is very important, and also a source of much frustration both for laboratory chemists and computational chemists. To begin with, the chemical reactions that produce chiral molecules often don't distinguish between the forms, producing both chiralities in equal amounts. (These products are called *racemic mixtures*.) So if you want to end up with just one form, your manufacturing process immediately becomes more complicated. In addition, many physical properties are identical for both chiralities, so many experiments can't distinguish between chiral versions of a given molecule. The same is true of computational models. For example, both chiralities have identical molecular graphs, so any machine learning model that depends only on the molecular graph will be unable to distinguish between them.

This wouldn't matter so much if the two forms behaved identically in practice, but that often is not the case. It is possible for the two chiral forms of a drug to bind to totally different proteins, and to have very different effects in your body. In many cases, only one form of a drug has the desired therapeutic effect. The other form just produces extra side effects without having any benefit.

One specific example of the differing effects of chiral compounds is the drug thalidomide, which was prescribed as a sedative in the 1950s and 1960s. This drug was subsequently available over the counter as a treatment for nausea and morning sickness associated with pregnancy. The R form of thalidomide is an effective sedative, while the S form is teratogenic and has been shown to cause severe birth defects. These difficulties are further compounded by the fact that thalidomide interconverts, or racemizes, between the two different forms in the body.

Featurizing a Molecule

With these descriptions of basic chemistry in hand, how do we get started with featurizing molecules? In order to perform machine learning on molecules, we need to transform them into feature vectors that can be used as inputs to models. In this section, we will discuss the DeepChem featurization submodule dc.feat, and explain how to use it to featurize molecules in a variety of fashions.

SMILES Strings and RDKit

SMILES is a popular method for specifying molecules with text strings. The acronym stands for "Simplified Molecular-Input Line-Entry System", which is sufficiently awkward-sounding that someone must have worked hard to come up with it. A SMILES string describes the atoms and bonds of a molecule in a way that is both concise and reasonably intuitive to chemists. To nonchemists, these strings tend to look like meaningless patterns of random characters. For example, "OCCc1c(C)[n+] (cs1)Cc2cnc(C)nc2N" describes the important nutrient thiamine, also known as vitamin B1.

DeepChem uses SMILES strings as its format for representing molecules inside datasets. There are some deep learning models that directly accept SMILES strings as their inputs, attempting to learn to identify meaningful features in the text representation. But much more often, we first convert the string into a different representation (or featurize it) better suited to the problem at hand.

DeepChem depends on another open source chemoinformatics package, RDKit, to facilitate its handling of molecules. RDKit provides lots of features for working with SMILES strings. It plays a central role in converting the strings in datasets to molecular graphs and the other representations described below.

Extended-Connectivity Fingerprints

Chemical fingerprints are vectors of 1s and 0s that represent the presence or absence of specific features in a molecule. Extended-connectivity fingerprints (ECFPs) are a class of featurizations that combine several useful features. They take molecules of arbitrary size and convert them into fixed-length vectors. This is important because lots of models require their inputs to all have exactly the same size. ECFPs let you take molecules of many different sizes and use them all with the same model. ECFPs are also very easy to compare. You can simply take the fingerprints for two molecules and compare corresponding elements. The more elements that match, the more similar the molecules are. Finally, ECFPs are fast to compute.

Each element of the fingerprint vector indicates the presence or absence of a particular molecular feature, defined by some local arrangement of atoms. The algorithm begins by considering every atom independently and looking at a few properties of the atom: its element, the number of covalent bonds it forms, etc. Each unique combination of these properties is a feature, and the corresponding elements of the vector are set to 1 to indicate their presence. The algorithm then works outward, combining each atom with all the ones it is bonded to. This defines a new set of larger features, and the corresponding elements of the vector are set. The most common variant of this technique is the ECFP4 algorithm, which allows for sub-fragments to have a radius of two bonds around a central atom.

The RDKit library provides utilities for computing ECFP4 fingerprints for molecules. DeepChem provides convenient wrappers around these functions. The dc.feat.CircularFingerprint class inherits from Featurizer and provides a standard interface to featurize molecules:

```
smiles = ['C1CCCCC1', 'O1CCOCC1'] # cyclohexane and dioxane
mols = [Chem.MolFromSmiles(smile) for smile in smiles]
feat = dc.feat.CircularFingerprint(size=1024)
arr = feat.featurize(mols)
# arr is a 2-by-1024 array containing the fingerprints for
# the two molecules
```

ECFPs do have one important disadvantage: the fingerprint encodes a large amount of information about the molecule, but some information does get lost. It is possible for two different molecules to have identical fingerprints, and given a fingerprint, it is impossible to uniquely determine what molecule it came from.

Molecular Descriptors

An alternative line of thought holds that it's useful to describe molecules with a set of physiochemical descriptors. These usually correspond to various computed quantities that describe the molecule's structure. These quantities, such as the log partition coefficient or the polar surface area, are often derived from classical physics or chemistry. The RDKit package computes many such physical descriptors on molecules. The DeepChem featurizer dc.feat.RDKitDescriptors() provides a simple way to perform the same computations:

```
feat = dc.feat.RDKitDescriptors()
arr = feat.featurize(mols)
# arr is a 2-by-200 array containing properties of the
# two molecules
```

This featurization is obviously more useful for some problems than others. It will tend to work best for predicting things that depend on relatively generic properties of the molecules. It is unlikely to work for predicting properties that depend on the detailed arrangement of atoms.

Graph Convolutions

The featurizations described in the preceding section were designed by humans. An expert thought carefully about how to represent molecules in a way that could be used as input to machine learning models, then coded the representation by hand. Can we instead let the model figure out for itself the best way to represent molecules?

That is what machine learning is all about, after all: instead of designing a featurization ourselves, we can try to learn one automatically from the data.

As an analogy, consider a convolutional neural network for image recognition. The input to the network is the raw image. It consists of a vector of numbers for each pixel, for example the three color components. This is a very simple, totally generic representation of the image. The first convolutional layer learns to recognize simple patterns such as vertical or horizontal lines. Its output is again a vector of numbers for each pixel, but now it is represented in a more abstract way. Each number represents the presence of some local geometric feature.

The network continues through a series of layers. Each one outputs a new representation of the image that is more abstract than the previous layer's representation, and less closely connected to the raw color components. And these representations are automatically learned from the data, not designed by a human. No one tells the model what patterns to look for to identify whether the image contains a cat. The model figures that out by itself through training.

Graph convolutional networks take this same idea and apply it to graphs. Just as a regular CNN begins with a vector of numbers for each pixel, a graph convolutional network begins with a vector of numbers for each node and/or edge. When the graph represents a molecule, those numbers could be high-level chemical properties of each atom, such as its element, charge, and hybridization state. Just as a regular convolutional layer computes a new vector for each pixel based on a local region of its input, a graph convolutional layer computes a new vector for each node and/or edge. The output is computed by applying a learned convolutional kernel to each local region of the graph, where "local" is now defined in terms of edges between nodes. For example, it might compute an output vector for each atom based on the input vector for that same atom and any other atoms it is directly bonded to.

That is the general idea. When it comes to the details, many different variations have been proposed. Fortunately, DeepChem includes implementations of lots of those architectures, so you can try them out even without understanding all the details. Examples include graph convolutions (`GraphConvModel`), Weave models (`WeaveModel`), message passing neural networks (`MPNNModel`), deep tensor neural networks (`DTNNModel`), and more.

Graph convolutional networks are a powerful tool for analyzing molecules, but they have one important limitation: the calculation is based solely on the molecular graph. They receive no information about the molecule's conformation, so they cannot hope to predict anything that is conformation-dependent. This makes them most suitable for small, mostly rigid molecules. In the next chapter we will discuss methods that are more appropriate for large, flexible molecules that can take on many conformations.

Training a Model to Predict Solubility

Let's put all the pieces together and train a model on a real chemical dataset to predict an important molecular property. First we'll load the data:

```
tasks, datasets, transformers = dc.molnet.load_delaney(featurizer='GraphConv')
train_dataset, valid_dataset, test_dataset = datasets
```

This dataset contains information about solubility, which is a measure of how easily a molecule dissolves in water. This property is vitally important for any chemical you hope to use as a drug. If it does not dissolve easily, getting enough of it into a patient's bloodstream to have a therapeutic effect may be impossible. Medicinal chemists spend a lot of time modifying molecules to try to increase their solubility.

Notice that we specify the option `featurizer='GraphConv'`. We are going to use a graph convolutional model, and this tells MoleculeNet to transform the SMILES string for each molecule into the format required by the model.

Now let's construct and train the model:

```
model = GraphConvModel(n_tasks=1, mode='regression', dropout=0.2)
model.fit(train_dataset, nb_epoch=100)
```

We specify that there is only one task—that is to say, one output value (the solubility) —for each sample. We also specify that this is a regression model, meaning that the labels are continuous numbers and the model should try to reproduce them as accurately as possible. That is in contrast to a classification model, which tries to predict which of a fixed set of classes each sample belongs to. To reduce overfitting, we specify a dropout rate of 0.2, meaning that 20% of the outputs from each convolutional layer will randomly be set to 0.

That's all there is to it! Now we can evaluate the model and see how well it works. We will use the Pearson correlation coefficient as our evaluation metric:

```
metric = dc.metrics.Metric(dc.metrics.pearson_r2_score)
print(model.evaluate(train_dataset, [metric], transformers))
print(model.evaluate(test_dataset, [metric], transformers))
```

This reports a correlation coefficient of 0.91 for the training set, and 0.70 for the test set. Apparently it is overfitting a little bit, but not too badly. And a correlation coefficient of 0.70 is quite respectable. Our model is successfully predicting the solubilities of molecules based on their molecular structures!

Now that we have the model, we can use it to predict the solubilities of new molecules. Suppose we are interested in the following five molecules, specified as SMILES strings:

```
smiles = ['COC(C)(C)CCCC(C)CC=CC(C)=CC(=O)OC(C)C',
          'CCOC(=O)CC',
          'CSc1nc(NC(C)C)nc(NC(C)C)n1',
```

```
'CC(C#C)N(C)C(=O)Nc1ccc(Cl)cc1',
'Cc1cc2ccccc2cc1C']
```

To use these as inputs to the model, we must first use RDKit to parse the SMILES strings, then use a DeepChem featurizer to convert them to the format expected by the graph convolution:

```
from rdkit import Chem
mols = [Chem.MolFromSmiles(s) for s in smiles]
featurizer = dc.feat.ConvMolFeaturizer()
x = featurizer.featurize(mols)
```

Now we can pass them to the model and ask it to predict their solubilities:

```
predicted_solubility = model.predict_on_batch(x)
```

MoleculeNet

We have now seen two datasets loaded from the `molnet` module: the Tox21 toxicity dataset in the previous chapter, and the Delaney solubility dataset in this chapter. MoleculeNet is a large collection of datasets useful for molecular machine learning. As shown in Figure 4-10, it contains data on many sorts of molecular properties. They range from low-level physical properties that can be calculated with quantum mechanics up to very high-level information about interactions with a human body, such as toxicity and side effects.

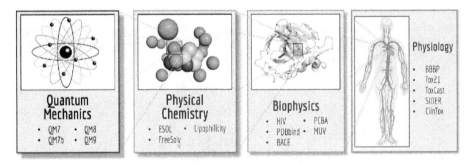

Figure 4-10. MoleculeNet hosts many different datasets from different molecular sciences. Scientists find it useful to predict quantum, physical chemistry, biophysical, and physiological characteristics of molecules.

When developing new machine learning methods, you can use MoleculeNet as a collection of standard benchmarks to test your method on. At *http://moleculenet.ai* you can view data on how well a collection of standard models perform on each of the datasets, giving insight into how your own method compares to established techniques.

SMARTS Strings

In many commonly used applications, such as word processing, we need to search for a particular text string. In cheminformatics, we encounter similar situations where we want to determine whether atoms in a molecule match a particular pattern. There are a number of use cases where this may arise:

- Searching a database of molecules to identify molecules containing a particular substructure
- Aligning a set of molecules on a common substructure to improve visualization
- Highlighting a substructure in a plot
- Constraining a substructure during a calculation

SMARTS is an extension of the SMILES language described previously that can be used to create queries. One can think of SMARTS patterns as similar to regular expressions used for searching text. For instance, when searching a filesystem, one can specify a query like "foo*.bar", which will match foo.bar, foo3.bar, and foolish.bar. At the simplest level, any SMILES string can also be a SMARTS string. The SMILES string "CCC" is also a valid SMARTS string and will match sequences of three adjacent aliphatic carbon atoms. Let's take a look at a code example showing how we can define molecules from SMILES strings, display those molecules, and highlight the atoms matching a SMARTS pattern.

First, we will import the necessary libraries and create a list of molecules from a list of SMILES strings. Figure 4-11 shows the result:

```
from rdkit import Chem
from rdkit.Chem.Draw import MolsToGridImage

smiles_list = ["CCCCC","CCOCC","CCNCC","CCSCC"]
mol_list = [Chem.MolFromSmiles(x) for x in smiles_list]
```

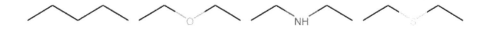

Figure 4-11. Chemical structures generated from SMILES

Now we can see which SMILES strings match the SMARTS pattern "CCC" (Figure 4-12):

```
query = Chem.MolFromSmarts("CCC")
match_list = [mol.GetSubstructMatch(query) for mol in
mol_list]
MolsToGridImage(mols=mol_list, molsPerRow=4,
highlightAtomLists=match_list)
```

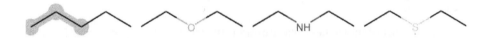

Figure 4-12. Molecules matching the SMARTS expression "CCC."

There are a couple of things to note in this figure. The first is that the SMARTS expression only matches the first structure. The other structures do not contain three adjacent carbons. Note also that there are multiple ways that the SMARTS pattern could match the first molecule in this figure—it could match three adjacent carbon atoms by starting at the first, second, or third carbon atom. There are additional functions in RDKit that will return all possible SMARTS matches, but we won't cover those now.

Additional wildcard characters can be used to match specific sets of atoms. As with text, the "*" character can be used to match any atom. The SMARTS pattern "C*C" will match an aliphatic carbon attached to any atom attached to another aliphatic carbon (see Figure 4-13).

```
query = Chem.MolFromSmarts("C*C")
match_list = [mol.GetSubstructMatch(query) for mol in
mol_list]
MolsToGridImage(mols=mol_list, molsPerRow=4,
highlightAtomLists=match_list)
```

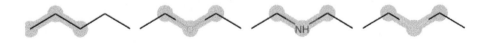

*Figure 4-13. Molecules matching the SMARTS expression "C*C".*

The SMARTS syntax can be extended to only allow specific sets of atoms. For instance, the string "C[C,O,N]C" will match carbon attached to carbon, oxygen, or nitrogen, attached to another carbon (Figure 4-14):

```
query = Chem.MolFromSmarts("C[C,N,O]C")
match_list = [mol.GetSubstructMatch(query) for mol in
mol_list]
MolsToGridImage(mols=mol_list, molsPerRow=4,
highlightAtomLists=match_list)
```

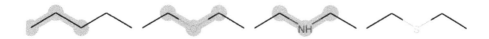

Figure 4-14. Molecules matching the SMARTS expression "C[C,N,O]C".

There is a lot more to SMARTS that is beyond the scope of this brief introduction. Interested readers are urged to read the "Daylight Theory Manual" to get deeper insight into SMILES and SMARTS.[1] As we will see in Chapter 11, SMARTS can be used to build up sophisticated queries that can identify molecules that may be problematic in biological assays.

Conclusion

In this chapter, you've learned the basics of molecular machine learning. After a brief review of basic chemistry, we explored how molecules have traditionally been represented for computing systems. You also learned about graph convolutions, which are a newer approach to modeling molecules in deep learning, and saw a complete working example of how to use machine learning on molecules to predict an important physical property. These techniques will serve as the foundations upon which later chapters will build.

---

1 Daylight Chemical Information Systems, Inc. "Daylight Theory Manual." *http://www.daylight.com/dayhtml/doc/theory/*. 2011.

## CHAPTER 5

Biophysical Machine Learning

In this chapter, we will explore how to use deep learning for understanding biophysical systems. In particular, we will explore in depth the problem of predicting how small drug-like molecules bind to a protein of interest in the human body.

This problem is of fundamental interest in drug discovery. Modulating a single protein in a targeted fashion can often have a significant therapeutic impact. The breakthrough cancer drug Imatinib tightly binds with BCR-ABL, for example, which is part of the reason for its efficacy. For other diseases, it can be challenging to find a single protein target with the same efficacy, but the abstraction remains useful nevertheless. There are so many mechanisms at play in the human body that finding an effective mental model can be crucial.

Drugs Don't Just Target a Single Protein

As we've discussed, it can be extraordinarily useful to reduce the problem of designing a drug for a disease to the problem of designing a drug that interacts tightly with a given protein. But it's extremely important to realize that in reality, any given drug is going to interact with many different subsystems in the body. The study of such multifaceted interactions is broadly called polypharmacology.

At present, computational methods for dealing with polypharmacology are still relatively undeveloped, so the gold standard for testing for polypharmacological effects remains animal and human experimentation. As computational techniques mature, this state of affairs may shift over the next few years.

Our goal therefore is to design learning algorithms that can effectively predict when a given molecule is going to interact with a given protein. How can we do this? For starters, we might borrow some of the techniques from the previous chapter on molecular machine learning and try to create a protein-specific model. Such a model would, given a dataset of molecules that either bind or don't bind to a given protein, learn to predict for new molecules whether they bind or not. This idea isn't actually terrible, but requires a good amount of data for the system at hand. Ideally, we'd have an algorithm that could work without a large amount of data for a new protein.

The trick, it turns out, is to use the physics of the protein. As we will discuss in the next section, quite a bit is known about the physical structure of protein molecules. In particular, it's possible to create snapshots of the 3D state of proteins using modern experimental techniques. These 3D snapshots can be fed into learning algorithms and used to predict binding. It's also possible to take snapshots (better known as *structures*) of the interaction of proteins with smaller molecules (often called *ligands*). If this discussion seems abstract to you right now, don't worry. You'll be seeing plenty of practical code in this chapter.

We'll begin this chapter with a deeper overview of proteins and their function in biology. We will then shift into computer science and introduce some algorithms for featurizing protein systems which can transform biophysical systems into vectors or tensors for use in learning. In the last part of the chapter, we will work through an in-depth case study on constructing a protein–ligand binding interaction model. For experimentation,we will introduce the PDBBind dataset, which contains a collection of experimentally determined protein–ligand structures. We will demonstrate how to featurize this dataset with DeepChem. We will then build some models, both deep and simpler, on these featurized datasets and study their performance.

Why Is It Called Biophysics?

It is often said that all of biology is based on chemistry, and all of chemistry is based on physics. At first glance, biology and physics may seem far removed from one another. But as we will discuss at greater length later in this chapter, physical laws are at the heart of all biological mechanisms. In addition, much of the study of protein structure depends critically on the use of experimental techniques refined in physics. Manipulating nanoscale machines (for that's what proteins actually are) requires considerable physical sophistication, on both the theoretical and the practical side.

It's also interesting to note that the deep learning algorithms we will discuss in this chapter bear significant similarities to deep learning architectures used for studying systems from particle physics or for physical simulations. Such topics are outside the scope of this book, but we encourage interested readers to explore them further.

Protein Structures

Proteins are tiny machines that do most of the work in a cell. Despite their small size, they can be very complicated. A typical protein is made of thousands of atoms arranged in precise ways.

To understand any machine, you must know what parts it is made of and how they are put together. You cannot hope to understand a car until you know it has wheels on the bottom, an empty space in the middle to hold passengers, and doors through which the passengers can enter and exit. The same is true of a protein. To understand how it works, you must know exactly how it is put together.

Furthermore, you need to know how it interacts with other molecules. Few machines operate in isolation. A car interacts with the passengers it carries, the road it drives on, and the energy source that allows it to move. This applies to most proteins as well. They act on molecules (for example, to catalyze a chemical reaction), are acted upon by others (for example, to regulate their activity), and draw energy from still others. All these interactions depend on the specific positioning of atoms in the two molecules. To understand them, you must know how the atoms are arranged in 3D space.

Unfortunately, you can't just look at a protein under a microscope; they are far too small for that. Instead, scientists have had to invent complex and ingenious methods for determining the structures of proteins. At present there are three such methods: *X-ray crystallography, nuclear magnetic resonance* (NMR for short), and *cryo-electron microscopy* (cryo-EM for short).

X-ray crystallography is the oldest method, and still the most widely used. Roughly 90% of all known protein structures were determined with this method. Crystallography involves growing a crystal of the protein of interest (many molecules of the protein all tightly packed together in a regular repeating pattern). X-rays are then shone on the crystal, the scattered light is measured, and the results are analyzed to work out the structure of the individual molecules. Despite its success, this method has many limitations. It is slow and expensive even in the best cases. Many proteins do not form crystals, making crystallography impossible. Packing the protein into a crystal may alter its structure, so the result might be different from its structure in a living cell. Many proteins are flexible and can take on a range of structures, but crystallography only produces a single unmoving snapshot. But even with these limitations, it is a remarkably powerful and important tool.

NMR is the second most common method. It operates on proteins in solution, so there is no need to grow a crystal. This makes it an important alternative for proteins that cannot be crystallized. Unlike crystallography, which produces a single fixed snapshot, NMR produces an ensemble of structures representing the range of shapes the protein can take on in solution. This is a very important benefit, since it gives information about how the protein can move. Unfortunately, NMR has its own limitations. It requires a highly concentrated solution, so it is mostly limited to small, highly soluble proteins.

In recent years, cryo-EM has emerged as a third option for determining protein structures. It involves rapidly freezing protein molecules, then imaging them with an electron microscope. Each image is far too low-resolution to make out the precise details; but by combining many different images, one can produce a final structure whose resolution is much higher than any individual electron microscope image. After decades of steady improvements to the methods and technologies, cryo-EM has finally begun to approach atomic resolution. Unlike crystallography and NMR, it works for large proteins that do not crystallize. This will probably make it a very important technique in the years to come.

The Protein Data Bank (PDB) (*https://www.rcsb.org/*) is the primary repository for known protein structures. At present it contains over 142,000 structures, like the one in Figure 5-1. That may seem like a lot, but it is far less than we really want. The number of known proteins is orders of magnitude larger, with more being discovered all the time. For any protein you want to study, there is a good chance that its structure is still unknown. And you really want many structures for each protein, not just one. Many proteins can exist in multiple functionally different states (for example, "active" and "inactive" states), so you want to know the structure of each state. If a protein binds to other molecules, you want a separate structure with the protein bound to each one so you can see exactly how they bind. The PDB is a fantastic resource, but the field as a whole is still in its "low data" stage. We have far less data than we want, and a major challenge is figuring out how to make the most of what we have. That is likely to remain true for decades.

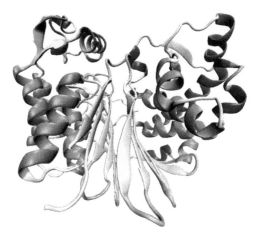

Figure 5-1. A crystal structure of the CapD protein from Bacillus anthracis, the anthrax pathogen. Determining the structures of bacterial proteins can be a powerful tool for antibiotic design. More generally, identifying the structure of a therapeutically relevant protein is one of the key steps in modern drug discovery.

Protein Sequences

So far in this chapter, we've primarily discussed protein structures, but we haven't yet said much about what proteins are made of atomically. Proteins are built out of fundamental building blocks called *amino acids*. These amino acids are sets of molecules that share a common core, but have different "side chains" attached (Figure 5-2). These different side chains alter the behavior of the protein.

A protein is a chain of amino acids linked one to the next to the next (Figure 5-3). The start of the amino acid chain is typically referred to as the N-terminus, while the end of the chain is called the C-terminus. Small chains of amino acids are commonly called *peptides*, while longer chains are called proteins. Peptides are too small to have complex 3D structures, but the structures of proteins can be very complicated.

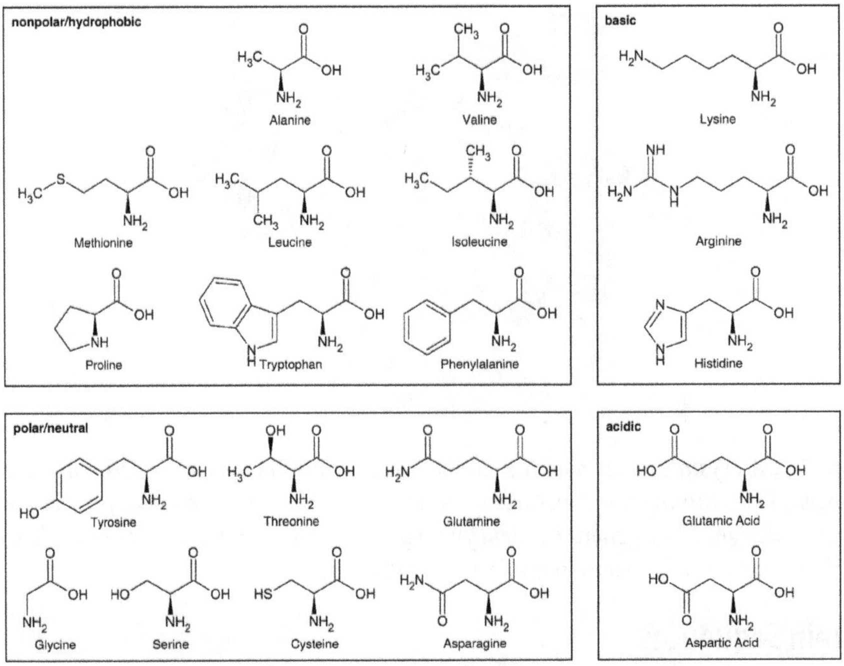

Figure 5-2. Amino acids are the building blocks of protein structures. This diagram represents the chemical structures of a number of commonly seen amino acids. (Adapted from Wikimedia (https://commons.wikimedia.org/wiki/File:Overview_proteinogenic_amino_acids-DE.svg).)

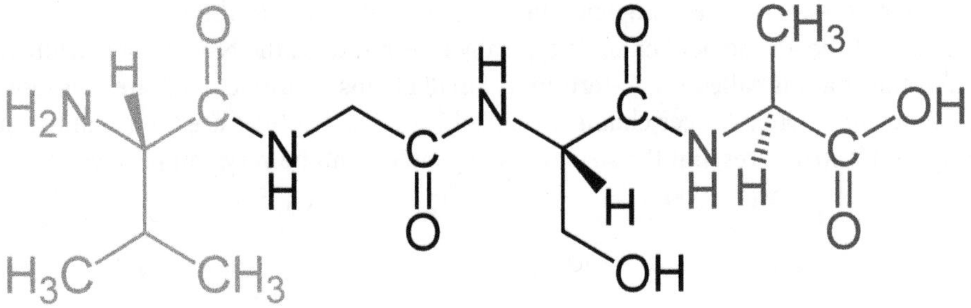

Figure 5-3. A chain of four amino acids, with the N-terminus on the left and the C-terminus on the right. (Source: Wikipedia (https://en.wikipedia.org/wiki/N-terminus#/media/File:Tetrapeptide_structural_formulae_v.1.png).)

It's worth noting that while most proteins take a rigid shape, there are also intrinsically disordered proteins which have regions that refuse to take rigid shapes (Figure 5-4).

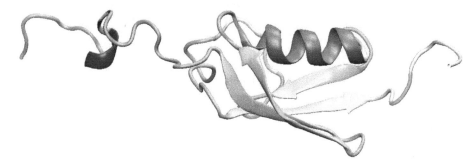

Figure 5-4. A snapshot of the SUMO-1 protein. The central core of the protein has struc-ture, while the N-terminal and C-terminal regions are disordered. Intrinsically disor-dered proteins such as SUMO-1 are challenging to handle computationally.

In the remainder of this chapter, we will primarily deal with proteins that have rigid, 3D shapes. Dealing with floppy structures with no set shape is still challenging for modern computational techniques.

Can't We Predict 3D Protein Structure Computationally?

After reading this section, you might wonder why we don't use algorithms to directly predict the structure of interesting protein molecules rather than depending on com-plex physical experiments. It's a good question, and there have in fact been decades of work on the computational prediction of protein structure.

There are two main approaches to predicting protein structures. The first is called *homology modeling*. Protein sequences and structures are the product of billions of years of evolution. If two proteins are near relatives (the technical term is "homo-logs") that only recently diverged from each other, they probably have similar struc-tures. To predict a protein's structure by homology modeling, you first look for a homolog whose structure is already known, then try to adjust it based on differences between the sequences of the two proteins. Homology modeling works reasonably well for determining the overall shape of a protein, but it often gets details wrong. And of course, it requires that you already know the structure of a homologous pro-tein.

The other main approach is *physical modeling*. Using knowledge of the laws of phys-ics, you try to explore many different conformations the protein might take on and predict which one will be most stable. This method requires enormous amounts of computing time. Until about a decade ago, it simply was impossible. Even today it is only practical for small, fast-folding proteins. Furthermore, it requires physical approximations to speed up the calculation, and those reduce the accuracy of the result. Physical modeling will often predict the right structure, but not always.

A Short Primer on Protein Binding

We've discussed a good amount about protein structure so far in this chapter, but we haven't said much about how proteins interact with other molecules (Figure 5-5). In practice, proteins often bind to small molecules. Sometimes that binding behavior is central to the protein's function: the main role for a given protein can involve binding to particular molecules. For example, signaling transduction in cells often passes messages via the mechanism of a protein binding to another molecule. Other times, the molecule binding to the protein is foreign: possibly a drug we've created to manipulate the protein, possibly a toxin that interferes with its function.

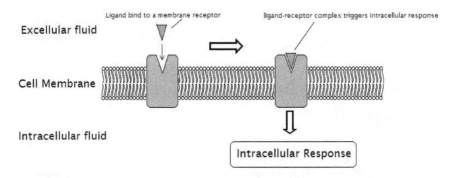

Figure 5-5. A signal transduced via a protein embedded in a cell's membrane. (Source: Wikimedia (https://simple.wikipedia.org/wiki/Signal_transduction#/media/ File:The_External_Reactions_and_the_Internal_Reactions.jpg).)

Understanding the details of how, where, and when molecules bind to proteins is critical to understanding their functions and developing drugs. If we can coopt the signaling mechanisms of cells in the human body, we can induce desired medical responses in the body.

Protein binding involves lots of very specific interactions, which makes it hard to predict computationally. A tiny change in the positions of just a few atoms can determine whether or not a molecule binds to a protein. Furthermore, many proteins are flexible and constantly moving. A protein might be able to bind a molecule when it's in certain conformations, but not when it's in others. Binding in turn may cause further changes to a protein's conformation, and thus to its function.

In the remainder of this chapter, we will use the challenge of understanding protein binding as a motivating computational example. We will delve in depth into current deep learning and machine learning approaches for making predictions about binding events.

Biophysical Featurizations

As we discussed in the previous chapter, one of the crucial steps in applying machine learning to a new domain is figuring out how to transform (or featurize) training data to a format suitable for learning algorithms. We've discussed a number of techniques for featurizing individual small molecules. Could we perhaps adapt these techniques for use in biophysical systems?

Unfortunately, the behaviors of biophysical systems are critically constrained by their 3D structures, so the 2D techniques from previous chapters miss crucial information. As a result, we will discuss a pair of new featurization techniques in this chapter. The first featurization technique, the *grid featurization*, explicitly searches a 3D structure for the presence of critical physical interactions such as hydrogen bonds and salt bridges (more on these later), which are known to play an important role in determining protein structure. The advantage of this technique is that we can rely upon a wealth of known facts about protein physics. The weakness, of course, is that we are bound by known physics and lessen the chance that our algorithms will be able to detect new physics.

The alternative featurization technique is the *atomic featurization*, which simply provides a processed representation of the 3D positions and identities of all atoms in the system. This makes the challenge for the learning algorithm considerably harder, since it must learn to identify critical physical interactions, but it also makes it feasible for learning algorithms to detect new patterns of interesting behavior.

PDB Files and Their Pitfalls

Protein structures are often stored in PDB files. Such files are simply text files that contain descriptions of the atoms in the structure and their positions in coordinate space relative to one another. Featurization algorithms typically rely on libraries that read in PDB files and store them into in-memory data structures. So far so good, right?

Unfortunately, PDB files are often malformed. The reason lies in the underlying physics. Often, an experiment will fail to have adequate resolution to completely specify a portion of the protein's structure. Such regions are left unspecified in the PDB file, so it's common to find that many atoms or even entire substructures of the protein are missing from the core structure.

Libraries such as DeepChem will often attempt to do the "right" thing and algorithmically fill in such missing regions. It's important to note that this cleanup is only approximate, and there's still no entirely satisfactory replacement to having an expert human peer at the protein structure (in a suitable viewing program) and point out issues. Hopefully, software tooling to handle these errors will improve over the next few years and the need for expert guidance will lessen.

Grid Featurization

By converting biophysical structures into vectors, we can use machine learning algorithms to make predictions about them. It stands to reason that it would be useful to have a featurization algorithm for processing protein–ligand systems. However, it's quite nontrivial to devise such an algorithm. Ideally, a featurization technique would need to have significant knowledge about the chemistry of such systems baked into it by design, so it could pull out useful features.

These features might include, for example, counts of noncovalent bonds between the protein and ligand, such as hydrogen bonds or other interactions. (Most protein–ligand systems don't have covalent bonds between the protein and ligand.)

Luckily for us, DeepChem has such a featurizer available. Its `RdkitGridFeaturizer` summarizes a set of relevant chemical information into a brief vector for use in learning algorithms. While it's not necessary to understand the underlying science in depth to use the featurizer, it will still be useful to have a basic understanding of the underlying physics. So, before we dive into a description of what the grid featurizer computes, we will first review some of the pertinent biophysics of macromolecular complexes.

While reading this section, it may be useful to refer back to the discussion of basic chemical interactions in the previous chapter. Ideas such as covalent and noncovalent bonds will pop up quite a bit.

The grid featurizer searches for the presence of such chemical interactions within a given structure and constructs a feature vector that contains counts of these interactions. We will say more about how this is done algorithmically later in the chapter.

Hydrogen bonds

When a hydrogen atom is covalently bonded to a more electronegative atom such as oxygen or nitrogen, the shared electrons spend most of their time closer to the more electronegative atom. This leaves the hydrogen with a net positive charge. If that positively charged hydrogen then gets close to another atom with a net negative charge, they are attracted to each other. That is a hydrogen bond (Figure 5-6).

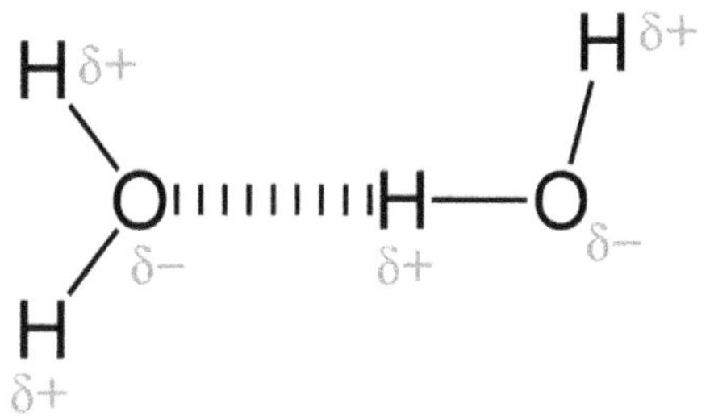

Figure 5-6. A rendered example of a hydrogen bond. Excess negative charge on the oxygen interacts with excess positive charge on the hydrogen, creating a bonding interaction. (Source: Wikimedia (https://commons.wikimedia.org/wiki/File:Hydrogen-bonding-in-water-2D.png).)

Because hydrogen atoms are so small, they can get very close to other atoms, leading to a strong electrostatic attraction. This makes hydrogen bonds one of the strongest noncovalent interactions. They are a critical form of interaction that often stabilizes molecular systems. For example, water's unique properties are due in large part to the network of hydrogen bonds that form between water molecules.

The RdkitGridFeaturizer attempts to count the hydrogen bonds present in a structure by checking for pairs of protein/ligand atoms of the right types that are suitably close to one another. This requires applying a cutoff to the distance, which is somewhat arbitrary. In reality there is not a sharp division between atoms being bonded and not bonded. This may lead to some misidentified interactions, but empirically, a simple cutoff tends to work reasonably well.

Salt bridges

A salt bridge is a noncovalent attraction between two amino acids, where one has a positive charge and the other has a negative charge (see Figure 5-7). It combines both ionic bonding and hydrogen bonding. Although these bonds are relatively weak, they can help stabilize the structure of a protein by providing an interaction between distant amino acids in the protein's sequence.

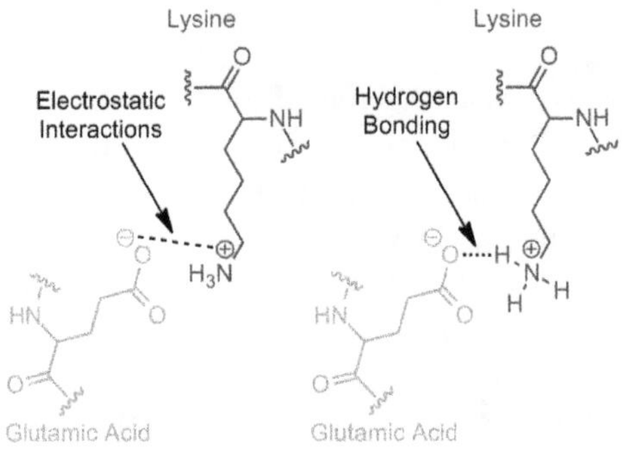

Figure 5-7. An illustration of a salt bridge between glutamic acid and lysine. The salt bridge is a combination of an ionic-style electrostatic interaction and a hydrogen bond and serves to stabilize the structure. (Source: Wikimedia (https://commons.wikimedia.org/wiki/File:Revisited_Glutamic_Acid_Lysine_salt_bridge.png).)

The grid featurizer attempts to detect salt bridges by explicitly checking for pairs of amino acids (such as glutamic acid and lysine) that are known to form such interactions, and that are in close physical proximity in the 3D structure of the protein.

Pi-stacking interactions

Pi-stacking interactions are a form of noncovalent interaction between *aromatic rings* (Figure 5-8). These are flat, ring-shaped structures that appear in many biological molecules, including DNA and RNA. They also appear in the side chains of some amino acids, including phenylalanine, tyrosine, and tryptophan.

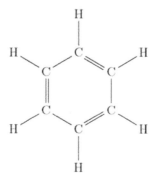

Figure 5-8. An aromatic ring in the benzene molecule. Such ring structures are known for their exceptional stability. In addition, aromatic rings have all their atoms lying in a plane. Heterogeneous rings, in contrast, don't have their atoms occupying the same plane.

Roughly speaking, pi-stacking interactions occur when two aromatic rings "stack" on top of each other. Figure 5-9 shows some of the ways in which two benzene rings can interact. Such stacking interactions, like salt bridges, can help stabilize various macro-molecular structures. Importantly, pi-stacking interactions can be found in ligand-protein interactions, since aromatic rings are often found in small molecules. The grid featurizer counts these interactions by detecting the presence of aromatic rings and checking for the distances between their centroids and the angles between their two planes.

Sandwich   Edge to Face   Displaced

Figure 5-9. Various noncovalent aromatic ring interactions. In the displaced interaction, the centers of the two aromatic rings are slightly displaced from one another. In edge-to-face interactions, one aromatic ring's edge stacks on another's face. The sandwich config-uration has two rings stacked directly, but is less energetically favorable than displaced or edge-to-face interactions since regions with the same charge interact closely.

At this point, you might be wondering why this type of interaction is called pi-stacking. The name refers to pi-bonds, a form of covalent chemical bond where the electron orbitals of two covalently bonded atoms overlap. In an aromatic ring, all the atoms in the ring participate in a joint pi-bond. This joint bond accounts for the stability of the aromatic ring and also explains many of its unique chemical properties.

For those readers who aren't chemists, don't worry too much if this material doesn't make too much sense just yet. DeepChem abstracts away many of these implementation details, so you won't need to worry much about pi-stacking on a regular basis when developing. However, it is useful to know that these interactions exist and play a major role in the underlying chemistry.

Intricate Geometries and Snapshots

In this section, we've introduced a number of interactions in terms of static geometric configurations. It's very important to realize that bonds are dynamic entities, and that in real physical systems, bonds will stretch, snap, break, and reform with dizzying speed. Keep this in mind, and note that when someone says a salt bridge exists, what they really mean is that in some statistically average sense, a salt bridge is likely present more often than not at a particular location.

Fingerprints

From the previous chapter, you may recall the use of circular fingerprints. These fingerprints count the number of fragments of a given type in the molecule, then use a hash function to fit these fragment counts into a fixed-length vector. Such fragment counts can be used for 3D molecular complexes as well. Although merely counting the fragments is often insufficient to compute the geometry of the system, the knowledge of present fragments can nevertheless be useful for machine learning systems. This might perhaps be due to the fact that the presence of certain fragments can be strongly indicative of some molecular events.

Some implementation details

To search for chemical features such as hydrogen bonds, the dc.feat.RdkitGridFea turizer needs to be able to effectively work with the geometry of the molecule. DeepChem uses the RDKit library to load each molecule, protein, and ligand, into a common in-memory object. These molecules are then transformed into NumPy arrays that contain the positions of all the atoms in space. For example, a molecule with N atoms can be represented as a NumPy array of shape $(N, 3)$, where each row represents the position of an atom in 3D space.

Then, performing a (crude) detection of a hydrogen bond simply requires looking at all pairs of atoms that could conceivably form a hydrogen bond (such as oxygen and hydrogen) that are sufficiently close to one another. The same computational strategy is used for detecting other kinds of bonds. For handling aromatic structures, there's a bit of special code to detect the presence of aromatic rings in the structure and compute their centroids.

Atomic Featurization

At the end of the previous section, we gave a brief overview of how features such as hydrogen bonds are computed by the RdkitGridFeaturizer. Most operations transform a molecule with N atoms into a NumPy array of shape (N, 3) and then perform a variety of extra computations starting from these arrays.

You can easily imagine that featurization for a given molecule could simply involve computing this (N, 3) array and passing it to a suitable machine learning algorithm. The model could then learn for itself what features were important, rather than relying on a human to select them and code them by hand.

In fact, this turns out to work—with a couple of extra steps. The (N, 3) position array doesn't distinguish atom types, so you also need to provide another array that lists the atomic number of each atom. As a second implementation-driven note, computing pairwise distances between two position arrays of shape (N, 3) can be very computationally expensive. It's useful to create "neighbor lists" in a preprocessing step, where the neighbor list maintains a list of neighboring atoms close to any given atom.

DeepChem provides a dc.feat.NeighborListAtomicCoordinates featurizer that handles much of this for you. We will not discuss it further in this chapter, but it's good to know that it exists as another option.

The PDBBind Case Study

With this introduction in place, let's start tinkering with some code samples for handling biophysical datasets. We will start by introducing the PDBBind dataset and the problem of binding free energy prediction. We will then provide code examples of how to featurize the PDBBind dataset and demonstrate how to build machine learning models for it. We will end the case study with a discussion of how to evaluate the results.

PDBBind Dataset

The PDBBind dataset contains a large number of biomolecular crystal structures and their binding affinities. There's a bit of jargon there, so let's stop and unpack it. A biomolecule is any molecule of biological interest. That includes not just proteins, but also nucleic acids (such as DNA and RNA), lipids, and smaller drug-like molecules. Much of the richness of biomolecular systems results from the interactions of various biomolecules with one another (as we've discussed at length already). A binding affinity is the experimentally measured affinity of two molecules to form a complex, with the two molecules interacting. If it is energetically favorable to form such a complex, the molecules will spend more time in that configuration as opposed to another one.

The PDBBind dataset has gathered structures of a number of biomolecular complexes. The large majority of these are protein–ligand complexes, but the dataset also contains protein–protein, protein–nucleic acid, and nucleic acid–ligand complexes. For our purposes, we will focus on the protein–ligand subset. The full dataset contains close to 15,000 such complexes, with the "refined" and "core" sets containing smaller but cleaner subsets of complexes. Each complex is annotated with an experimental measurement of the binding affinity for the complex. The learning challenge for the PDBBind dataset is to predict the binding affinity for a complex given the protein–ligand structure.

The data for PDBBind is gathered from the Protein Data Bank. Note that the data in the PDB (and consequently PDBBind) is highly heterogeneous! Different research groups have different experimental setups, and there can be high experimental variance between different measurements by different groups. For this reason, we will primarily use the filtered refined subset of the PDBBind dataset for doing our experimental work.

Dynamics Matter!

In this case study, we treat the protein and ligand as a frozen snapshot. Note that this is very unphysical! The protein and ligand are actually in rapid movement, and the ligand will move in and out of the protein's binding pocket. In addition, the protein may not even have one fixed binding pocket. For some proteins, there are a number of different sites where potential ligands interact.

All these combinations of factors mean that our models will have relatively limited accuracy. If we had more data, it might be possible that strong learning models could learn to account for these factors, but with more limited datasets, it is challenging to do so.

You should probably note this information. The design of better biophysical deep learning models to accurately account for the thermodynamic behavior of these systems remains a major open problem.

What If You Don't Have a Structure?

Drug discovery veterans might pause for a minute here. The fact is that it's typically much harder experimentally to determine the structure of a complex than it is to measure a binding affinity. This makes sense intuitively. A binding affinity is a single number for a given biomolecular complex, while a structure is a rich 3D snapshot. Predicting the binding affinity from the structure might feel a little bit like putting the cart before the horse!

There are a couple of answers to this (fair) accusation. The first is that the problem of determining the binding affinity of a biomolecular system is a physically interesting problem in its own right. Checking that we can accurately predict such binding affinities is a worthy test problem to benchmark our machine learning methods and can serve as a stepping stone to designing deep architectures capable of understanding sophisticated biophysical systems.

The second answer is that we can use existing computational tooling, such as "docking" software, to predict approximate structures for a protein–ligand complex given that we have a structure for the protein in isolation. While predicting a protein structure directly is a formidable challenge, it's somewhat easier to predict the structure of a protein–ligand complex when you already have the protein structure. So, you might well be able to make useful predictions with the system we will create in this case study by pairing it with a docking engine. Indeed, DeepChem has support for this use case, but we will not delve into this more advanced feature in this chapter.

In general, when doing machine learning, it can be particularly useful to take a look at individual data points or files within a dataset. In the code repository associated with this book (*https://github.com/deepchem/DeepLearningLifeSciences*), we've included the PDB file for a protein–ligand complex called 2D3U. It contains information about the amino acids (also called residues) of the protein.In addition, the PDB file contains the coordinates of each atom in 3D space. The units for these coordinates are angstroms (1 angstrom is 10^{-10} meters). The origin of this coordinate system is arbitrarily set to help in visualizing the protein, and is often set at the centroid of the protein. We recommend taking a minute to open this file in a text editor and take a look.

Why Is an Amino Acid Called a Residue?

As you spend more time working with biophysical data, you will commonly come across the terminology of an amino acid being called a *residue*. This refers to the chemistry of how proteins are formed. When two amino acids are joined together in a growing chain, an oxygen and two hydrogens are removed. A "residue" is what remains of an amino acid after this reaction takes place.

It can be very hard to understand the contents of a PDB file, so let's visualize a protein. We will use the NGLview (*https://github.com/arose/nglview*) visualization package, which integrates well with Jupyter notebooks. In the notebook associated with this chapter in the code repository, you will be able to manipulate and interact with the visualized protein. For now, Figure 5-10 shows a visualization of a protein–ligand complex (2D3U) generated within the Jupyter notebook.

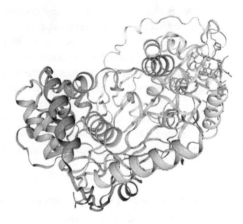

Figure 5-10. A visualization of the 2D3U protein–ligand complex from the PDBBind dataset. Note that the protein is represented in cartoon format for ease of visualization, and that the ligand (near the top-right corner) is represented in ball-and-stick format for full detail.

Notice how the ligand rests in a sort of "pocket" in the protein. You can see this more clearly by rotating the visualization to look at it from different sides.

Protein Visualization Tools

Given the importance of visualizing proteins to work with them, there are a number of protein visualization tools available. While NGLview has amazing Jupyter integration, it's more common to see other tools, such as VMD (*https://www.ks.uiuc.edu/Research/vmd/*), PyMOL (*https://pymol.org*), or Chimera (*https://www.cgl.ucsf.edu/chimera/*), in use by professional drug discoverers. Note, however, that these tools are often not fully open source, and may not feature a developer-friendly API. Nevertheless, if you plan to spend significant time working with protein structures, using one of these more established tools is probably still worth the trade-off.

Featurizing the PDBBind Dataset

Let's start by building a RdkitGridFeaturizerobject that we can inspect:

```
import deepchem as dc
featurizer = dc.feat.RdkitGridFeaturizer(
        voxel_width=2.0, sanitize=True, flatten=True,
        feature_types=['hbond', 'salt_bridge', 'pi_stack',
                        'cation_pi', 'ecfp', 'splif'])
```

There are a number of options here, so let's pause and consider what they mean. The sanitize=True flag asks the featurizer to try to clean up any structures it is given. Recall from our earlier discussion that structures are often malformed. The sanitization step will attempt to fix any obvious errors that it detects. Setting flatten=True asks the featurizer to output a one-dimensional feature vector for each input structure.

The feature_types flag sets the types of biophysical and chemical features that the RdkitGridFeaturizer will attempt to detect in input structures. Note the presence of many of the chemical bonds we discussed earlier in the chapter: hydrogen bonds, salt bridges, etc. Finally, the option voxel_width=2.0 sets the size of the voxels making up the grid to 2 angstroms. The RdkitGridFeaturizer converts a protein to a voxelized representation for use in extracting useful features. For each spatial voxel, it counts biophysical features and also computes a local fingerprint vector. The Rdkit GridFeaturizer computes two different types of fingerprints, the ECFP and SPLIF fingerprints.

Voxelization

What is voxelization? Broadly put, a voxel is the 3D analogue of a pixel (see Figure 5-11). Just as pixelized representations of images can be extraordinarily useful for handling imaging data, voxelized representations can be critical when working with 3D data.

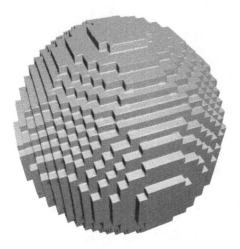

Figure 5-11. A voxelized representation of a sphere. Note how each voxel represents a spatial cube of input.

We are now ready to load the PDBBind dataset:

```
tasks, datasets, transformers = dc.molnet.load_pdbbind(
        featurizer=featurizer, splitter="random", subset="core")
train_dataset, valid_dataset, test_dataset = datasets
```

With this snippet, we've loaded and featurized the core subset of PDBBind. On a fast computer, this should run within a few minutes. Featurizing the refined subset will take a couple of hours on a modern server.

Now that we have the data in hand, let's start building some machine learning models. We'll first train a classical model called a *random forest*:

```
from sklearn.ensemble import RandomForestRegressor
sklearn_model = RandomForestRegressor(n_estimators=100)
model = dc.models.SklearnModel(sklearn_model)
model.fit(train_dataset)
```

As an alternative, we will also try building a neural network for predicting protein-ligand binding. We can use the dc.models.MultitaskRegressor class to build an MLP with two hidden layers. We set the widths of the hidden layers to 2,000 and 1,000, respectively, and use 50% dropout to reduce overfitting.:

```
n_features = train_dataset.X.shape[1]
model = dc.models.MultitaskRegressor(
        n_tasks=len(tasks),
        n_features=n_features,
        layer_sizes=[2000, 1000],
        dropouts=0.5,
```

```
        learning_rate=0.0003)
model.fit(train_dataset, nb_epoch=50)
```

 Baseline Models

Deep learning models are tricky to optimize correctly at times. It's
easy for even experienced practitioners to make errors when tuning
a deep model. For this reason, it's critical to construct a baseline
model that is more robust, even if it perhaps has lower performance.

Random forests are very useful choices for baselines. They often
offer strong performance on learning tasks with relatively small
amounts of tuning. A random forest classifier constructs many
"decision tree" classifiers, each using only a random subset of the
available features, then combines the individual decisions of these
classifiers via a majority vote.

Scikit-learn is an invaluable package for constructing simple
machine learning baselines. We will use scikit-learn to construct a
baseline model in this chapter, using the RdkitGridFeaturizer to
featurize complexes as inputs to random forests.

Now that we have a trained model, we can proceed to checking its accuracy. In order
to evaluate the accuracy of the model, we have to first define a suitable metric. Let's
use the Pearson R^2 score. This is a number between –1 and 1, where 0 indicates no
correlation between the true and predicted labels, and 1 indicates perfect correlation:

```
metric = dc.metrics.Metric(dc.metrics.pearson_r2_score)
```

Let's now evaluate the accuracy of the models on the training and test datasets
according to this metric. The code to do so is again shared between both models:

```
print("Evaluating model")
train_scores = model.evaluate(train_dataset, [metric], transformers)
test_scores = model.evaluate(test_dataset, [metric], transformers)

print("Train scores")
print(train_scores)

print("Test scores")
print(test_scores)
```

Many Architectures Can Have Similar Effects

In this section, we provide code examples of how to use an MLP with grid featurization to model protein–ligand structures in Deep-Chem. It's important to note that there are a number of alternative deep architectures that have similar effects. There's been a line of work on using 3D convolutional networks to predict protein–ligand binding interactions using voxel-based featurizations. Other work has used variants of the graph convolutions we saw in the previous chapter to handle macromolecular complexes.

What are the differences between these architectures? So far, it looks like most of them have similar predictive power. We use grid featurization because there's a tuned implementation in Deep-Chem, but other models may serve your needs as well. Future versions of DeepChem will likely include additional architectures for this purpose.

For the random forest, this reports a training set score of 0.979 but a test set score of only 0.133. It does an excellent job of reproducing the training data but a very poor job of predicting the test data. Apparently it is overfitting quite badly.

In comparison, the neural network has a training set score of 0.990 and a test set score of 0.359. It does slightly better on the training set and much better on the test set. There clearly is still overfitting going on, but the amount is reduced and the overall ability of the model to predict new data is much higher.

Knowing the correlation coefficient is a powerful first step toward understanding the model we've built, but it's always useful to directly visualize how our predictions correlate with actual experimental data. Figure 5-12 shows the true versus predicted labels for each of the models when run on the test set. We immediately see how the neural network's predictions are much more closely correlated with the true data than are the random forest's.

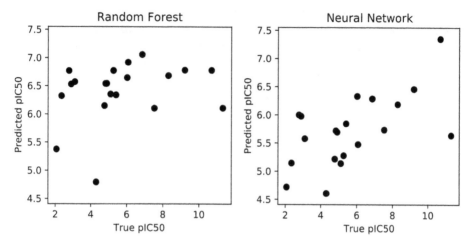

Figure 5-12. True versus predicted labels for the two models when run on the test set.

Conclusion

In this chapter you've learned about applying deep learning to biophysical systems, and in particular to the problem of predicting the binding affinity of protein–ligand systems. You might be curious how general the skills you've learned are. Could you apply these same models and techniques to understand other biophysical datasets? Let's do a quick survey and tour.

Protein–protein and protein–DNA systems follow the same basic physics as protein–ligand systems at a high level. The same hydrogen bonds, salt bridges, pi-stacking interactions, and so on play a critical role. Could we just reuse the code from this chapter to analyze such systems? The answer turns out to be a little complicated. Many of the physical interactions that drive protein–ligand interactions are driven by charged dynamics. Protein–protein dynamics may, on the other hand, be driven more by bulk hydrophobic interactions. We won't dig deeply into the meaning of these interactions, but they have a different character qualitatively than protein–ligand interactions to some degree. This could mean that RdkitGridFeaturizer wouldn't do a good job of characterizing these interactions. On the other hand, it's possible that the atomic convolutional models might do a better job of handling these systems, since much less about the physics of interactions is hardcoded into these deep models.

That said, there remains a significant problem of scale. The atomic convolutional models are quite slow to train and require a great deal of memory. Scaling these models to handle larger protein–protein systems would require additional work on the engineering end. The DeepChem development team is hard at work on these and other challenges, but more time may be required before these efforts reach fruition.

Antibody–antigen interactions are another form of critical biophysical interaction. Antibodies are Y-shaped proteins (see Figure 5-13) that have a variable "antigen-binding site" used to bind antigens. Here, antigens are molecules associated with a particular pathogen. Cells grown in culture can be harnessed to create antibodies that target specific antigens. If all the cells in a culture are clones of a given cell, then the antibodies produced will be identical. Such "monoclonal antibodies" have recently found wide therapeutic use.

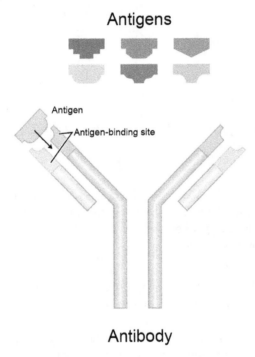

Figure 5-13. Diagram of antibody–antigen interaction. (Source: Wikimedia (https://en.wikipedia.org/wiki/Antibody#/media/File:Antibody.svg).)

The design of antibodies has primarily been an experimental science until now. Part of this is due to the challenges involved in getting a 3D antibody structure. However, modeling the complex antigen–antibody binding site has also proven a challenge. Some of the techniques we've covered in this chapter may find fruitful use in antibody–antigen binding modeling over the next few years.

We've also alluded to the importance of dynamics in understanding protein physics. Could we not do deep learning directly on protein simulations to understand which ligands could bind to the protein? In principle yes, but formidable challenges remain. Some companies are actively working on this challenge, but good open source tooling is not yet available.

In Chapter 11, we will return to some biophysical techniques and show how these models can be very useful for drug discovery work.

## CHAPTER 6

Deep Learning for Genomics

At the heart of every living organism is its genome: the molecules of DNA containing all the instructions to make the organism's working parts. If a cell is a computer, then its genome sequence is the software it executes. And if DNA can be seen as software, information meant to be processed by a computer, surely we can use our own computers to analyze that information and understand how it functions?

But of course, DNA is not just an abstract storage medium. It is a physical molecule that behaves in complicated ways. It also interacts with thousands of other molecules, all of which play important roles in maintaining, copying, directing, and carrying out the instructions contained in the DNA. The genome is a huge and complex machine made up of thousands of parts. We still have only a poor understanding of how most of those parts work, to say nothing of how they all come together as a working whole.

This brings us to the twin fields of *genetics* and *genomics*. Genetics treats DNA as abstract information. It looks at patterns of inheritance, or seeks correlations across populations, to discover the connections between DNA sequences and physical traits. Genomics, on the other hand, views the genome as a physical machine. It tries to understand the pieces that make up that machine and the ways they work together. The two approaches are complementary, and deep learning can be a powerful tool for both of them.

DNA, RNA, and Proteins

Even if you are not a biologist, at some point in your education you probably studied the basics of how genomes operate. We will first review the simplified picture of genomics that is usually taught in introductory classes. Then we will describe some of the ways in which the real world is more complicated.

DNA is a polymer: a long chain of repeating units strung together. In the case of DNA, there are four units (called *bases*) that can appear: adenine, cytosine, guanine, and thymine, which are abbreviated as A, C, G, and T (see Figure 6-1). Nearly all the information about how to make a living organism is ultimately encoded in the specific pattern of these four repeating units that make up its genome.

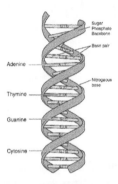

Figure 6-1. Structure of a DNA molecule. It consists of two chains, each made of many A, C, G, and T bases. The two chains are complementary: every C in one chain is paired with a G in the other, and every A in one chain is paired with a T in the other. (Source: Wikimedia (https://en.wikipedia.org/wiki/Molecular_Struc ture_of_Nucleic_Acids:_A_Structure_for_Deoxyribose_Nucleic_Acid#/media/ File:DNA-structure-and-bases.png).)

If DNA is the software, proteins are the most important hardware. Proteins are tiny machines that do almost all the work in a cell. Proteins are also polymers, made up of repeating units called *amino acids*. There are 20 main amino acids, and their physical properties vary widely. Some are large while others are small. Some have an electric charge while others do not. Some tend to attract water while others tend to repel it. When just the right set of amino acids is strung together in just the right order, it will spontaneously fold up into a 3D shape, all the pieces positioned just right to let it function as a machine.

One of the main functions of DNA is to record the sequences of amino acids for an organism's proteins. It does this in a simple, straightforward way. Particular stretches of DNA directly correspond to particular proteins. Each sequence of three DNA bases (called a *codon*) corresponds to one amino acid. For example, the pattern AAA indicates the amino acid lysine, while the pattern GCC indicates the amino acid alanine.

Going from DNA to protein involves another molecule, RNA, that serves as an intermediate representation to carry information from one part of the cell to another. RNA is yet another polymer and is chemically very similar to DNA. It too has four bases that can be chained together in arbitrary orders. To create a protein, the information must be copied twice. First the DNA sequence is *transcribed* into an equivalent RNA sequence, and then the RNA molecule is *translated* into a protein molecule. The RNA molecule that carries the information is called a *messenger RNA*, or mRNA for short.

This tells us *how* proteins get made, but not *when*. A human cell has many thousands of different proteins it can make. Surely it doesn't just churn out copies of all of them, all the time? Clearly there must be some sort of regulatory mechanism to control which proteins get made when. In the conventional picture, this is done by special proteins called *transcription factors (TFs)*. Each TF recognizes and binds to a particular DNA sequence. Depending on the particular TF and the location where it binds, it can either increase or decrease the rate at which nearby genes are transcribed.

This gives a simple, easy-to-understand picture of how a genome works. The job of DNA is to encode proteins. Stretches of DNA (called *genes*) code for proteins using a simple, well-defined code. DNA is converted to RNA, which serves only as an information carrier. The RNA is then converted into proteins, which do all the real work. The whole process is very elegant, the sort of thing a talented engineer might have designed. And for many years, this picture was believed to be mostly correct. So, take a moment to enjoy it before we spoil the view by revealing that reality is actually far messier and far more complicated.

And Now for the Real World

Now it's time to talk about how genomes *really* work. The picture described in the previous section is simple and elegant, but unfortunately it has little connection to reality. This section will go through a lot of information very quickly, but don't worry about remembering or understanding all of it. The important thing is just to get a sense of the incredible complexity of living organisms. We will return to some of these subjects later in the chapter and discuss them in more detail.

Let's begin by considering DNA molecules (called *chromosomes*). In bacteria, which have relatively small genomes, DNA exists as simple free-floating molecules. But eukaryotes (a group that includes amoebas, humans, and everything in between) have much larger genomes. To fit inside the cell, each chromosome must be packed into a very small space. This is accomplished by winding it around proteins called *histones*. But if all the DNA is tightly packed away, how can it be transcribed? The answer, of course, is that it can't. Before a gene can be transcribed, the stretch of DNA containing it first must be unwound. How does the cell know which DNA to unwind? The answer is still poorly understood. It is believed to involve various types of chemical modification to the histone molecules, and proteins that recognize particular modifications. Clearly there is a regulatory mechanism involved, but many of the details are still unknown. We will return to this subject shortly.

DNA itself can be chemically modified through a process called *methylation*. The more highly a stretch of DNA is methylated, the less likely it is to be transcribed, so this is another regulatory mechanism the cell can use to control the production of proteins. But how does it control which regions of DNA are methylated? This too is still poorly understood.

In the previous section we said that a particular stretch of DNA corresponds to a particular protein. That is correct for bacteria, but in eukaryotes the situation is more complicated. After the DNA is transcribed into a messenger RNA, that RNA often is edited to remove sections and connect (or *splice*) the remaining parts (called *exons*) back together again. The RNA sequence that finally gets translated into a protein may therefore be different from the original DNA sequence. In addition, many genes have multiple *splice variants*—different ways of removing sections to form the final sequence. This means a single stretch of DNA can actually code for several different proteins!

Is all of this starting to sound very complicated? Well, keep reading, because we've barely started! Evolution selects for mechanisms that work, without any concern for whether they are simple or easy to understand. It leads to very complicated systems, and understanding them requires us to confront that complexity.

In the conventional picture RNA is viewed as just an information carrier, but even from the early days of genomics, biologists knew that was not entirely correct. The job of translating mRNA to proteins is performed by *ribosomes*, complicated molecular machines made partly of proteins and partly of RNA. Another key role in translation is performed by molecules called *transfer RNAs* (or tRNAs for short). These are the molecules that define the "genetic code," recognizing patterns of three bases in mRNA and adding the correct amino acid to the growing protein. So, for over half a century we've known there were at least three kinds of RNA: mRNA, ribosomal RNA, and tRNA.

But RNA still had lots of tricks up its sleeve. It is a surprisingly versatile molecule. Over the last few decades, many other types of RNA have been discovered. Here are some examples:

- *Micro RNAs* (miRNAs) are short pieces of RNA that bind to a messenger RNA and prevent it from being translated into proteins. This is a very important regulatory mechanism in some types of animals, especially mammals.

- *Short interfering RNA* (siRNA) is another type of RNA that binds to mRNA and prevents it from being translated. It's similar to miRNA, but siRNAs are double stranded (unlike miRNAs, which are single stranded), and some of the details of how they function are different. We will discuss siRNA in more detail later in the chapter.

- *Ribozymes* are RNA molecules that can act as enzymes to catalyze chemical reactions. Chemistry is the foundation of everything that happens in a living cell, so catalysts are vital to life. Usually this job is done by proteins, but we now know it sometimes is done by RNA.
- *Riboswitches* are RNA molecules that consist of two parts. One part acts as a messenger RNA, while the other part is capable of binding to a small molecule. When it binds, that can either enable or prevent translation of the mRNA. This is yet another regulatory mechanism by which protein production can be adjusted based on the concentration of particular small molecules in the cell.

Of course, all these different types of RNA must be manufactured, and the DNA must contain instructions on how to make them. So, DNA is more than just a string of encoded protein sequences. It also contains RNA sequences, plus binding sites for transcription factors and other regulatory molecules, plus instructions for how messenger RNAs should be spliced, plus various chemical modifications that influence how it is wound around histones and which genes get transcribed.

Now consider what happens after the ribosome finishes translating the mRNA into a protein. Some proteins can spontaneously fold into the correct 3D shape, but many others require help from other proteins called *chaperones*. It is also very common for proteins to need additional chemical modifications after they are translated. Then the finished protein must be transported to the correct location in the cell to do its job, and finally degraded when it is no longer needed. Each of these processes is controlled by additional regulatory mechanisms, and involves interactions with lots of other molecules.

If this all sounds overwhelming, that's because it is! A living organism is far more complicated than any machine ever created by humans. The thought of trying to understand it *should* intimidate you!

But this is also why machine learning is such a powerful tool. We have huge amounts of data, generated by a process that is both mind-bogglingly complex and poorly understood. We want to discover subtle patterns buried in the data. This is exactly the sort of problem that deep learning excels at!

In fact, deep learning is *uniquely* well suited to the problem. Classical statistical techniques struggle to represent the complexity of the genome. They often are based around simplifying assumptions. For example, they look for linear relationships between variables, or they try to model a variable as depending on only a small number of other variables. But genomics involves complex nonlinear relationships between hundreds of variables: exactly the sort of relationship that can be effectively described by a deep neural network.

Transcription Factor Binding

As an example of applying deep learning to genomics, let's consider the problem of predicting transcription factor binding. Recall that TFs are proteins that bind to DNA. When they bind, they influence the probability of nearby genes being transcribed into RNA. But how does a TF know where to bind? Like so much of genomics, this question has a simple answer followed by lots of complications.

To a first approximation, every TF has a specific DNA sequence called its *binding site motif* that it binds to. Binding site motifs tend to be short, usually 10 bases or less. Wherever a TF's motif appears in the genome, the TF will bind to it.

In practice, though, motifs are not completely specific. A TF may be able to bind to many similar but not identical sequences. Some bases within the motif may be more important than others. This is often modeled as a *position weight matrix* that specifies how much preference the TF has for each possible base at each position within the motif. Of course, that assumes every position within the motif is independent, which is not always true. Sometimes even the length of a motif can vary. And although binding is primarily determined by the bases within the motif, the DNA to either side of it can also have some influence.

And that's just considering the sequence! Other aspects of the DNA can also be important. Many TFs are influenced by the physical shape of the DNA, such as how tightly the double helix is twisted. If the DNA is methylated, that can influence TF binding. And remember that most DNA in eukaryotes is tightly packed away, wound around histones. TFs can only bind to the portions that have been unwound.

Other molecules also play important roles. TFs often interact with other molecules, and those interactions can affect DNA binding. For example, a TF may bind to a second molecule to form a complex, and that complex then binds to a different DNA motif than the TF would on its own.

Biologists have spent decades untangling these details and designing models for TF binding. Instead of doing that, let's see if we can use deep learning to learn a model directly from data.

A Convolutional Model for TF Binding

For this example, we will use experimental data on a particular transcription factor called JUND. An experiment was done to identify every place in the human genome where it binds. To keep things manageable, we only include the data from chromosome 22, one of the smallest human chromosomes. It is still over 50 million bases long, so that gives us a reasonable amount of data to work with. The full chromosome has been split up into short segments, each 101 bases long, and each segment has been labeled to indicate whether it does or does not include a site where JUND binds.

We will try to train a model that predicts those labels based on the sequence of each segment.

The sequences are represented with one-hot encoding. For each base we have four numbers, of which one is set to 1 and the others are set to 0. Which of the four numbers is set to 1 indicates whether the base is an A, C, G, or T.

To process the data we will use a convolutional neural network, just like we did for recognizing handwritten digits in Chapter 3. In fact, you will see the two models are remarkably similar to each other. This time we will use 1D convolutions, since we are dealing with 1D data (DNA sequences) instead of 2D data (images), but the basic components of the model will be the same: inputs, a series of convolutional layers, one or more dense layers to compute the output, and a cross entropy loss function.

Let's start by creating a layer to define the inputs:

```
features = tf.keras.Input(shape=(101, 4))
```

Notice the shape of the input. For each sample, we have a feature vector of size 101 (the number of bases) by 4 (the one-hot encoding of each base).

Next we create a stack of three convolutional layers, all with identical parameters:

```
import tensorflow.keras.layers as layers
prev = features
for i in range(3):
    prev = layers.Conv1D(filters=15, kernel_size=10,
                         activation=tf.nn.relu,
                         padding='same')(prev)
    prev = layers.Dropout(rate=0.5)(prev)
```

We specify 10 for the width of the convolutional kernels, and that each layer should include 15 filters (that is, outputs). The first layer takes the raw features (four numbers per base) as input. It looks at spans of 10 consecutive bases, so 40 input values in total. For each span, it multiplies those 40 values by a convolutional kernel to produce 15 output values. The second layer again looks at spans of 10 bases, but this time the inputs are the 15 values computed by the first layer. It computes a new set of 15 values for each base, and so on.

To prevent overfitting, we add a dropout layer after each convolutional layer. The dropout probability is set to 0.5, meaning that 50% of all output values are randomly set to 0.

Next we use a dense layer to compute the output:

```
logits = layers.Dense(units=1)(layers.Flatten()(prev))
output = layers.Activation(tf.math.sigmoid)(logits)
keras_model = tf.keras.Model(inputs=features, outputs=[output, logits])
```

We want the output to be between 0 and 1 so we can interpret it as the probability a particular sample contains a binding site. The dense layer can produce arbitrary values, not limited to any particular range. We therefore pass it through a logistic sigmoid function to compress it to the desired range. The input to this function is often referred to as *logits*. The name refers to the mathematical logit function, which is the inverse function of the logistic sigmoid.

Finally, we create a `KerasModel`, telling it to use the cross entropy as the loss function:

```
model = dc.models.KerasModel(
    keras_model,
    loss=dc.models.losses.SigmoidCrossEntropy(),
    output_types=['prediction', 'loss'],
    batch_size=1000)
```

Notice that for reasons of numerical stability, the cross entropy loss is computed from logits instead of the output of the sigmoid function. We indicate this by specifying 'loss' as the output type for the second output (it will be passed to the loss function in place of the output used for making predictions).

Now we are ready to train and evaluate the model. We use ROC AUC as our evaluation metric. After every 10 epochs of training, we evaluate the model on both the training and validation sets:

```
train = dc.data.DiskDataset('train_dataset')
valid = dc.data.DiskDataset('valid_dataset')
metric = dc.metrics.Metric(dc.metrics.roc_auc_score)
for i in range(20):
    model.fit(train, nb_epoch=10)
    print(model.evaluate(train, [metric]))
    print(model.evaluate(valid, [metric]))
```

The result is shown in Figure 6-2. The validation set performance peaks at about 0.75 after 50 epochs, then decreases slightly. The training set performance continues to increase, eventually leveling off at around 0.87. This tells us that training beyond 50 epochs just leads to overfitting, and we should halt training at that point:

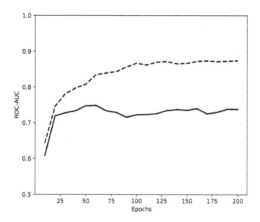

Figure 6-2. Evolution of ROC AUC scores during training for the training set (dashed) and validation set (solid).

A ROC AUC score of 0.75 is not bad, but also not wonderful. Possibly we could increase it by improving the model. There are lots of hyperparameters we could try changing: the number of convolutional layers, the kernel width for each layer, the number of filters in each layer, the dropout rate, etc. We could try lots of combinations for them, and we might find one with better performance.

But we also know there are fundamental limits to how well this model can ever work. The only input it looks at is the DNA sequence, and TF binding also depends on lots of other factors: accessibility, methylation, shape, the presence of other molecules, etc. Any model that ignores those factors will be limited in how accurate its predictions can ever be. So now let's try adding a second input and see if it helps.

Chromatin Accessibility

The name *chromatin* refers to everything that makes up a chromosome: DNA, histones, and various other proteins and RNA molecules. *Chromatin accessibility* refers to how accessible each part of the chromosome is to outside molecules. When the DNA is tightly wound around histones, is becomes inaccessible to transcription factors and other molecules. They cannot reach it, and the DNA is effectively inactive. When it unwinds from the histones, it becomes accessible again and resumes its role as a central part of the cell's machinery.

Chromatin accessibility is neither uniform nor static. It varies between types of cells and stages of a cell's life cycle. It can be affected by environmental conditions. It is one of the tools a cell uses to regulate the activity of its genome. Any gene can be turned off by packing away the area of the chromosome where it is located.

Accessibility also is constantly changing as DNA winds and unwinds in response to events in the cell. Instead of thinking of accessibility as a binary choice (accessible or inaccessible), it is better to think of it as a continuous variable (what fraction of the time each region is accessible).

The data we analyzed in the last section came from experiments on a particular kind of cell called HepG2. The experiments identified locations in the genome where the transcription factor JUND was bound. The results were influenced by chromatin accessibility. If a particular region is almost always inaccessible in HepG2 cells, the experiment was very unlikely to find JUND bound there, even if the DNA sequence would otherwise be a perfect binding site. So, let's try incorporating accessibility into our model.

First let's load some data on accessibility. We have it in a text file where each line corresponds to one sample from our dataset (a 101-base stretch of chromosome 22). A line contains the sample ID followed by a number that measures how accessible that region tends to be in HepG2 cells. We'll load it into a Python dictionary:

```python
span_accessibility = {}
for line in open('accessibility.txt'):
  fields = line.split()
  span_accessibility[fields[0]] = float(fields[1])
```

Now to build the model. We will use almost exactly the same model as in the previous section with just two minor changes. First, we need a second feature input for the accessibility values. It has one number for each sample:

```python
accessibility = tf.keras.Input(shape=(1,))
```

Now we need to incorporate the accessibility value into the calculation. There are many ways we might do this. For the purposes of this example, we will use a particularly simple method. In the previous section, we flattened the output of the last convolution layer, then used it as the input to a dense layer that calculated the output.

```python
logits = layers.Dense(units=1)(layers.Flatten()(prev))
```

This time we will do the same thing, but also append the accessibility to the output of the convolution:

```python
prev = layers.Concatenate()([layers.Flatten()(prev), accessibility])
logits = layers.Dense(units=1)(prev)
```

That's all there is to the model! Now it's time to train it.

At this point we run into a difficulty: our model has two different Input layers! Up until now, our models have had exactly one Input layer. We trained them by calling fit(dataset), which automatically connected the dataset's X field to the feature input. But that clearly can't work when the model has more than one set of features.

This situation is handled by using a more advanced feature of DeepChem. Instead of passing a dataset to the model, we can write a Python generator function that iterates over batches. Each batch is represented by a tuple of the form (inputs, labels, weights), where each entry is a list of NumPy arrays to pass to the model or loss function:

```
def generate_batches(dataset, epochs):
  for epoch in range(epochs):
    for X, y, w, ids in dataset.iterbatches(batch_size=1000,
                                            pad_batches=True):
      yield ([X, np.array([span_accessibility[id] for id in ids])], [y], [w])
```

Notice how the dataset takes care of iterating through batches for us. It provides the data for each batch, from which we can construct whatever inputs the model requires.

Training and evaluation now proceed exactly as before. We use alternate forms of the methods that take a generator instead of a dataset:

```
for i in range(20):
    model.fit_generator(generate_batches(train, epochs=10))
    print(model.evaluate_generator(generate_batches(train, 1), [metric]))
    print(model.evaluate_generator(generate_batches(valid, 1), [metric]))
```

The result is shown in Figure 6-3. Both the training and validation set scores are improved compared to the model that ignored chromatin accessibility. ROC AUC score now reaches 0.91 for the training set and 0.80 for the validation set.

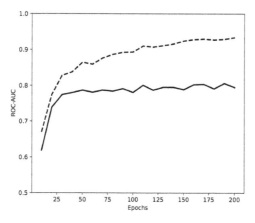

Figure 6-3. Evolution of ROC AUC scores during training for the training set (dashed) and validation set (solid) when including chromatin accessibility as an input.

RNA Interference

For our final example, let's turn to RNA. Much like DNA, this is a polymer composed of four repeating units called bases. In fact, three of the four bases are almost identical to their DNA versions, differing only in having one extra oxygen atom. The fourth base is a little more different. In place of thymine (T), RNA has a base called uracil (U). When a DNA sequence is transcribed into RNA, every T is replaced by a U.

The bases G and C are *complementary* to each other, in the sense that they have a strong tendency to bond to each other. Likewise, the bases A and T (or U) are complementary. If you have two strands of DNA or RNA, and every base in one is complementary to the corresponding base in the other, the two strands will tend to stick together. This fact plays a key role in lots of biological processes, including both transcription and translation, as well as DNA replication when a cell is dividing.

It also is central to something called *RNA interference*. This phenomenon was only discovered in the 1990s, and the discovery led to a Nobel Prize in 2006. A short piece of RNA whose sequence is complementary to part of a messenger RNA can bind to that mRNA. When this happens, it "silences" the mRNA and prevents it from being translated into a protein. The molecule that does the silencing is called a short interfering RNA (siRNA).

Except there is much more to the process than that. RNA interference is a complex biological mechanism, not just a side effect of two isolated RNA strands happening to stick together. It begins with the siRNA binding to a collection of proteins called the *RNA-induced silencing complex* (RISC). The RISC uses the siRNA as a template to search out matching mRNAs in the cell and degrade them. This serves both as a mechanism for regulating gene expression and as a defense against viruses.

It also is a powerful tool for biology and medicine. It lets you temporarily "turn off" any gene you want. You can use it to treat a disease, or to study what happens when a gene is disabled. Just identify the mRNA you want to block, select any short segment of it, and create a siRNA molecule with the complementary sequence.

Except that (of course!) it isn't as simple as that. You can't really just pick any segment of the mRNA at random, because (of course!) RNA molecules aren't just abstract patterns of four letters. They are physical objects with distinct properties, and those properties depend on the sequence. Some RNA molecules are more stable than others. Some bind their complementary sequences more strongly than others. Some fold into shapes that make it harder for the RISC to bind them. This means that some siRNA sequences work better than others, and if you want to use RNA interference as a tool, you need to know how to select a good one!

Biologists have developed lots of heuristics for selecting siRNA sequences. They will say, for example, that the very first base should be either an A or G, that G and C bases should make up between 30% and 50% of the sequence, and so on. These heuristics are helpful, but let's see if we can do better using machine learning.

We'll train our model using a library of 2,431 siRNA molecules, each 21 bases long.[1] Every one of them has been tested experimentally and labeled with a value between 0 and 1, indicating how effective it is at silencing its target gene. Small values indicate ineffective molecules, while larger values indicate more effective ones. The model takes the sequence as input and tries to predict the effectiveness.

Here is the code to build the model:

```
features = tf.keras.Input(shape=(21, 4))
prev = features
for i in range(2):
    prev = layers.Conv1D(filters=10, kernel_size=10,
                         activation=tf.nn.relu,
                         padding='same')(prev)
    prev = layers.Dropout(rate=0.3)(prev)
output = layers.Dense(units=1, activation=tf.math.sigmoid)(layers.Flatten()(prev))
keras_model = tf.keras.Model(inputs=features, outputs=output)
model = dc.models.KerasModel(
    keras_model,
    loss=dc.models.losses.L2Loss(),
    batch_size=1000)
```

This is very similar to the model we used for TF binding, with just a few differences. Because we are working with shorter sequences and training on less data, we have reduced the size of the model. There are only 2 convolutional layers, and 10 filters per layer instead of 15.

We also use a different loss function. The model for TF binding was a classification model. Every label was either 0 or 1, and we tried to predict which of those two discrete values it was. But this one is a regression model. The labels are continuous numbers, and the model tries to match them as closely as possible. We therefore use the L_2 distance as our loss function, which tries to minimize the mean squared difference between the true and predicted labels.

Here is the code to train the model:

```
train = dc.data.DiskDataset('train_siRNA')
valid = dc.data.DiskDataset('valid_siRNA')
metric = dc.metrics.Metric(dc.metrics.pearsonr, mode='regression')
for i in range(20):
    model.fit(train, nb_epoch=10)
```

1 Huesken, D., J. Lange, C. Mickanin, J. Weiler, F. Asselbergs, J. Warner, B. Meloon, S. Engel, A. Rosenberg, D. Cohen, M. Labow, M. Reinhardt, F. Natt, and J. Hall, "Design of a Genome-Wide siRNA Library Using an Artificial Neural Network." *Nature Biotechnology* 23:995–1001. 2005. *https://doi.org/10.1038/nbt1118.*

```
print(model.evaluate(train, [metric])['pearsonr'])
print(model.evaluate(valid, [metric])['pearsonr'])
```

For TF binding, we used ROC AUC as our evaluation metric, which measures how accurately the model can divide the data into two classes. That is appropriate for a classification problem, but it doesn't make sense for a regression problem, so instead we use the Pearson correlation coefficient. This is a number between −1 and 1, where 0 means the model provides no information at all and 1 means the model perfectly reproduces the experimental data.

The result is shown in Figure 6-4. The validation set score peaks at 0.65 after 50 epochs. The training set score continues to increase, but since there is no further improvement to the validation set score this is just overfitting. Given the simplicity of the model and the limited amount of training data, a correlation coefficient of 0.65 is quite good. More complex models trained on larger datasets do slightly better, but this is already very respectable performance.

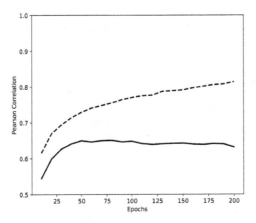

Figure 6-4. Evolution of the Pearson correlation coefficient during training for the training set (dashed) and validation set (solid).

Conclusion

A genome is an enormously complicated machine, with a vast number of parts all working together to direct and carry out the manufacture of proteins and other molecules. Deep learning is a powerful tool for studying it. Neural networks can pick out the subtle patterns in genomic data, providing insight into how the genome functions as well as making predictions about it.

Even more than most other areas of the life sciences, genomics produces huge amounts of experimental data. For example, a single human genome sequence includes more than six billion bases. Traditional statistical techniques struggle to find the signal buried in all that data. They often require simplifying assumptions that do not reflect the complexity of genomic regulation. Deep learning is uniquely suited to processing this data and advancing our understanding of the core functions of living cells.

## CHAPTER 7

Machine Learning for Microscopy

In this chapter, we introduce you to deep learning techniques for microscopy. In such applications, we seek to understand the biological structure of a microscopic image. For example, we might be interested in counting the number of cells of a particular type in a given image, or we might seek to identify particular organelles. Microscopy is one of the most fundamental tools for the life sciences, and advances in microscopy have greatly advanced human science. Seeing is believing even for skeptical scientists, and being able to visually inspect biological entities such as cells builds an intuitive understanding of the underlying mechanisms of life. A vibrant visualization of cell nuclei and cytoskeletons (as in Figure 7-1) builds a much deeper understanding than a dry discussion in a textbook.

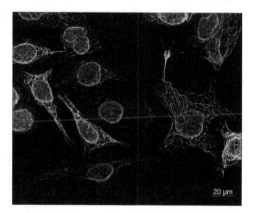

Figure 7-1. Human-derived SK8/18-2 cells. These cells are stained to highlight their nuclei and cytoskeletons and imaged using fluorescence microscopy. (Source: Wikimedia (https://commons.wikimedia.org/wiki/ File:SK8-18-2_human_derived_cells,_fluorescence_microscopy_(29942101073).jpg).)

The question remains how deep learning can make a difference in microscopy. Until recently, the only way to analyze microscopy images was to have humans (often graduate students or research associates) manually inspect these images for useful patterns. More recently, tools such as CellProfiler (*https://cellprofiler.org/*) have made it possible for biologists to automatically assemble pipelines for handling imaging data.

Automated High-Throughput Microscopy Image Analysis

Advances in automation in the last few decades have made it feasible to perform automated high-throughput microscopy on some systems. These systems use a combination of simple robotics (for automated handling of samples) and image processing algorithms to automatically process images. These image processing applications such as separating the foreground and background of cells and obtaining simple cell counts and other basic measurements. In addition, tools like CellProfiler have allowed biologists without programming experience to construct new automated pipelines for handling cellular data.

However, automated microscopy systems have traditionally faced a number of limitations. For one, complex visual tasks couldn't be performed by existing computer vision pipelines. In addition, properly preparing samples for analysis takes considerable sophistication on the part of the scientist running the experiment. For these reasons, automated microscopy has remained a relatively niche technique, despite its considerable success in enabling sophisticated new experiments.

Deep learning consequently holds considerable promise for extending the capabilities of tools such as CellProfiler. If deep analysis methods can perform more complex analyses, automated microscopy could become a considerably more effective tool. For this reason, there has been considerable research interest in deep microscopy, as we shall see in the remainder of this chapter.

The hope of deep learning techniques is that they will enable automated microscopy pipelines to become significantly more flexible. Deep learning systems show promise at being able to perform nearly any task a human image analyst can. In addition, early research suggests that deep learning techniques could considerably expand the capabilities of inexpensive microscopy hardware, potentially allowing cheap microscopes to perform analyses currently possible only using very sophisticated and expensive apparatuses.

Looking forward, it is even possible to train deep models that "simulate" experimental assays. Such systems are capable of predicting the outcomes of experiments (in some limited cases) without even running the experiment in question. This is a very powerful capability, and one which has spurred much excitement about the potential for deep networks in image-based biology.

In this chapter, we will teach you the basics of deep microscopy. We will demonstrate how deep learning systems can learn to perform simple tasks such as cell counting and cellular segmentation. In addition, we will discuss how to build extensible systems that could serve to handle more sophisticated image processing pipelines.

A Brief Introduction to Microscopy

Before we dive into algorithms, let's first talk basics. Microscopy is the science of using physical systems to view small objects. Traditionally, microscopes were purely optical devices, using finely ground lenses to expand the resolution of samples. More recently, the field of microscopy has started to lean heavily on technologies such as electron beams or even physical probes to produce high-resolution samples.

Microscopy has been tied intimately to the life sciences for centuries. In the 17th century, Anton van Leeuwenhoek used early optical microscopes (of his own design and construction) to describe microorganisms in unprecedented detail (as shown in Figure 7-2). These observations depended critically on van Leeuwenhoek's advances in microscopy, and in particular on his invention of a new lens which allowed for significantly improved resolution over the microscopes available at the time.

Figure 7-2. A reproduction of van Leeuwenhoek's microscope constructed in the modern era. Van Leeuwenhoek kept key details of his lens grinding process private, and a successful reproduction of the microscope wasn't achieved until the 1950s by scientists in the United States and the Soviet Union. (Source: Wikimedia (https://en.wikipedia.org/wiki/ Antonie_van_Leeuwenhoek#/media/File:Leeuwenhoek_Microscope.png).)

The invention of high-resolution optical microscopes triggered a revolution in microbiology. The spread of microscopy techniques and the ability to view cells, bacteria, and other microorganisms at scale enabled the entire field of microbiology and the pathogenic model of disease. It's hard to overstate the effect of microscopy on the modern life sciences.

Optical microscopes are either simple or compound. Simple microscopes use only a single lens for magnification. Compound microscopes use multiple lenses to achieve higher resolution, but at the cost of additional complexity in construction. The first practical compound microscopes weren't achieved until the middle of the 19th century! Arguably, the next major shift in optical microscopy system design didn't happen until the 1980s, with the advent of digital microscopes, which enabled the images captured by a microscope to be written to computer storage. As we mentioned in the previous section, automated microscopy uses digital microscopes to capture large volumes of images. These can be used to conduct large-scale biological experiments that capture the effects of experimental perturbations.

Modern Optical Microscopy

Despite the fact that optical microscopy has been around for centuries, there's still considerable innovation happening in the field. One of the most fundamental techniques is *optical sectioning*. An optical microscope has focal planes where the microscope is currently focused. A variety of techniques to focus the image on a chosen focal plane have been developed. These focused images can then be stitched together algorithmically to create a high-resolution image or even a 3D reconstruction of the original image. Figure 7-3 visually demonstrates how sectioned images of a grain of pollen can be combined to yield a high-fidelity image.

Confocal microscopes are a common solution to the problem of optical sectioning. They use a pinhole to block light coming in from out of focus, allowing a confocal microscope to achieve better depth perception. By shifting the focus of the microscope and doing a horizontal scan, you can get a full picture of the entire sample with increased optical resolution and contrast. In an interesting historical aside, the concept of confocal imaging was first patented by the AI pioneer Marvin Minsky (see Figure 7-4).

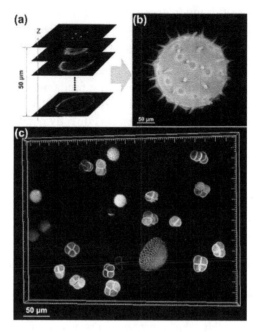

Figure 7-3. Pollen grain imaging: (a) optically sectioned fluorescence images of a pollen grain; (b) combined image; (c) combined image of a group of pollen grains. (Source: Wikimedia (https://commons.wikimedia.org/wiki/File:Optical_section ing_of_pollen.jpg).)

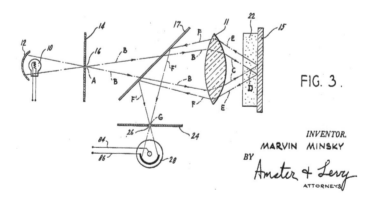

Figure 7-4. An image from Minsky's original patent introducing a confocal scanning microscope. In a curious twist of history, Minsky is better known for his pioneering work in AI. (Source: Wikimedia (https://en.wikipedia.org/wiki/Confocal_microscopy#/ media/File:Minsky_Confocal_Reflection_Microscope.png).)

Well-designed optical sectioning microscopes excel at capturing 3D images of biological systems since scans can be used to focus on multiple parts of the image. These focused images can be stitched together algorithmically, yielding beautiful 3D reconstructions.

In the next section, we will explore some of the fundamental limits that constrain optical microscopy and survey some of the techniques that have been designed to work around these limitations. This material isn't directly related to deep learning yet (for reasons we shall discuss), but we think it will give you a valuable understanding of the challenges facing microscopy today. This intuition will prove useful if you want to help design the next generation of machine learning–powered microscopy systems. However, if you're in a hurry to get to some code, we encourage you to skip forward to the subsequent sections where we dive into more immediate applications.

What Can Deep Learning Not Do?

It seems intuitively obvious that deep learning can make an impact in microscopy, since deep learning excels at image handling and microscopy is all about image capture. But it's worth asking: what parts of microscopy can't deep learning do much for now? As we see later in this chapter, preparing a sample for microscopic imaging can require considerable sophistication. In addition, sample preparation requires considerable physical dexterity, because the experimenter must be capable of fixing the sample as a physical object. How could we possibly automate or speed up this process with deep learning?

The unfortunate truth right now is that robotic systems are still very limited. While simple tasks like confocal scans of a sample are easy to handle, cleaning and preparing a sample requires considerable expertise. It's unlikely that any robotic systems available in the near future will have this ability.

Whenever you hear forecasts about the future impact of learning techniques, it's useful to keep examples like sample preparation in mind. Many of the pain points in the life sciences involve tasks such as sample preparation that just aren't feasible for today's machine learning. That may well change, but likely not for the next few years at the least.

The Diffraction Limit

When studying a new physical instrument such as a microscope, it can be useful to start by trying to understand its limits. What can't microscopes do? It turns out this question has been studied in great depth by previous generations of physicists (with some recent surprises too!). The first place to start is the *diffraction limit*, a theoretical limit on the resolution possible with a microscope:

$$d = \frac{\lambda}{2n \sin \theta}$$

The quantity $n \sin \theta$ is often rewritten as the numerical aperture, NA. λ is the wavelength of light. Note the implicit assumptions here. We assume that the sample is illuminated with some form of light. Let's take a quick look at the spectrum of light waves out there (see Figure 7-5).

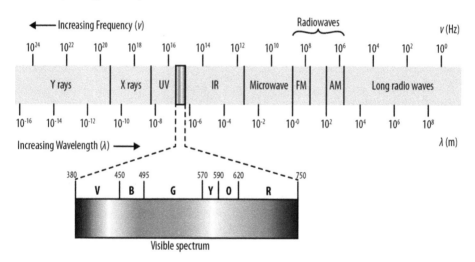

Figure 7-5. Wavelengths of light. Note that low-wavelength light sources such as X-rays are increasingly energetic. As a result, they will often destroy delicate biological samples.

Note how visible light forms only a tiny fraction of this spectrum. In principle, we should be able to make the desired resolution arbitrarily good using light at a low enough wavelength. To some extent this has happened already. A number of microscopes use electromagnetic waves of higher energy. For example, ultraviolet microscopes use the fact that UV rays have smaller wavelengths to allow for higher resolution. Couldn't we take this pattern further and use light of even smaller wavelength? For example, why not an X-ray or gamma-ray microscope? The main issue here is *phototoxicity*. Light with small wavelengths is highly energetic. Shining such light upon a sample can destroy the structure of the sample. In addition, high-wavelength light is dangerous for the experimenter and requires special experimental facilities.

Luckily, though, there exist a number of other techniques for bypassing the diffraction limit. One uses electrons (which have wavelengths too!) to image samples. Another uses physical probes instead of light. Yet another method for avoiding the resolution limit is to make use of near-field electromagnetic waves. Tricks with multiple illuminated fluorophores can also allow the limit to be lowered. We'll discuss these techniques in the following sections.

Electron and Atomic Force Microscopy

In the 1930s, the advent of the electron microscope triggered a dramatic leap in modern microscopy. The electron microscope uses electron beams instead of visible light in order to obtain images of objects. Since the wavelengths of electrons are much smaller than those of visible light, using electron beams instead of light waves allows much more detailed images. Why does this make any sense? Well, aren't electrons particles? Remember that matter can exhibit wave-like properties. This is known as the de Broglie wavelength, which was first proposed by Louis de Broglie:

$$\lambda = \frac{h}{p} = \frac{h}{mv}$$

Here, h is Planck's constant and m and v are the mass and velocity of the particle in question. (For the physicists, note that this formula doesn't account for relativistic effects. There are modified versions of the formula that do so.) Electron microscopes make use of the wave-like nature of electrons to image physical objects. The wavelength of an electron depends on its energy, but is easily subnanometer at wavelengths achievable by a standard electron gun. Plugging into the diffraction limit model discussed previously, it's easy to see how electron microscopy can be a powerful tool. The first prototype electron microscopes were constructed in the early 1930s. While these constructions have been considerably refined, today's electron microscopes still depend on the same core principles (see Figure 7-6).

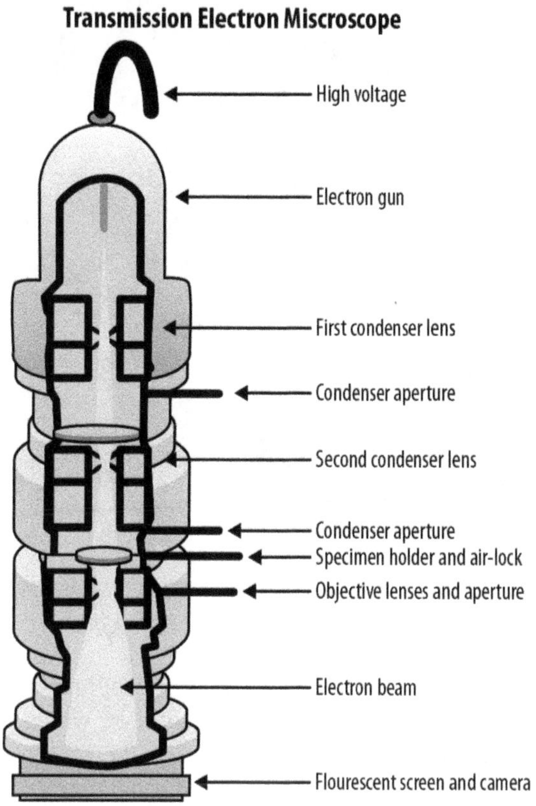

Transmission Electron Miscroscope

High voltage

Electron gun

First condenser lens

Condenser aperture

Second condenser lens

Condenser aperture
Specimen holder and air-lock
Objective lenses and aperture

Electron beam

Flourescent screen and camera

Figure 7-6. The components of a modern transmission electron microscope. (Source: Wikimedia (https://commons.wikimedia.org/wiki/File:Electron_Microscope.png).)

Note that we haven't entirely bypassed the issues with phototoxicity here. To get electrons with very small wavelengths, we need to increase their energy—and at very high energy, we will again destroy samples. In addition, the process of preparing samples for imaging by an electron microscope can be quite involved. Nevertheless, the use of electron microscopes has allowed for stunning images of microscopic systems (see Figure 7-7). Scanning electron microscopes, which scan the input sample to achieve larger fields of view,allow for images with resolution as small as one nanometer.

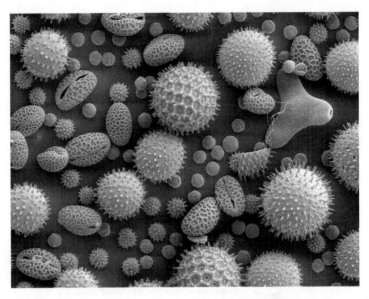

Figure 7-7. Pollen magnified 500x by a scanning electron microscope. (Source: Wikimedia (https://commons.wikimedia.org/wiki/File:Misc_pollen.jpg).)

Atomic force microscopy (AFM) provides another way of breaking through the optical diffraction limit. This technique leverages a cantilever which probes a given surface physically. The direct physical contact between the cantilever and the sample allows for pictures with resolutions at fractions of a nanometer. Indeed, it is sometimes possible to image single atoms! Atomic force microscopes also provide for 3D images of a surface due to the direct contact of the cantilever with the surface at hand.

Force microscopy broadly is a recent technique. The first atomic force microscopes were only invented in the 1980s, after nanoscale manufacturing techniques had matured to a point where the probes involved could be accurately made. As a result, applications in the life sciences are still emergent. There has been some work on imaging cells and biomolecules with AFM probes, but these techniques are still early.

Super-Resolution Microscopy

We've discussed a number of ways to stretch the diffraction limit so far in this chapter, including using higher-wavelength light or physical probes to allow for greater resolution. However, in the second half of the 20th century came a scientific breakthrough, led by the realization that there existed entire families of methods for breaking past the diffraction limit. Collectively, these techniques are called super-resolution microscopy techniques:

Functional super-resolution microscopy

Makes use of physical properties of light-emitting substances embedded in the sample being imaged. For example, fluorescent tags (more on these later) in biological microscopy can highlight particular biological molecules. These techniques allow standard optical microscopes to detect light emitters. Functional super-resolution techniques can be broadly split into deterministic and stochastic techniques.

Deterministic super-resolution microscopy

Some light-emitting substances have a nonlinear response to excitation. What does this actually mean? The idea is that arbitrary focus on a particular light emitter can be achieved by "turning off" the other emitters nearby. The physics behind this is a little tricky, but well-developed techniques such as stimulated emission depletion (STED) microscopy have demonstrated this technique.

Stochastic super-resolution microscopy

Light-emitting molecules in biological systems are subject to random motion. This means that if the motion of a light-emitting particle is tracked over time, its measurements can be averaged to yield a low error estimate of its true position. There are a number of techniques (such as STORM, PALM, and BALM microscopy) that refine this basic idea. These super-resolution techniques have had a dramatic effect in modern biology and chemistry because they allow relatively cheap optical equipment to probe the behavior of nanoscale systems. The 2014 Nobel Prize in Chemistry was awarded to pioneers of functional super-resolution techniques.

Deep Super-Resolution Techniques

Recent research has started to leverage the power of deep learning techniques to reconstruct super-resolution views.[1] These techniques claim orders of magnitude improvements in the speed of super-resolution microscopy by enabling reconstructions from sparse, rapidly acquired images. While still in its infancy, this shows promise as a future application area for deep learning.

Near-field microscopy is another super-resolution technique that makes use of local electromagnetic information in a sample. These "evanescent waves" don't obey the diffraction limit, so higher resolution is possible. However, the trade-off is that the microscope has to gather light from extremely close to the sample (within one wavelength of light from the sample). This means that although near-field techniques

---

1 Ouyang, Wei, et al. "Deep Learning Massively Accelerates Super-Resolution Localization Microscopy." *Nature Biotechnology* 36 (April 2018): 460–468. *https://doi.org/10.1038/nbt.4106.*

make for extremely interesting physics, practical use remains challenging. Very recently, it has also become possible to construct "metamaterials" which have a negative refractive index. In effect, the properties of these materials mean that near-field evanescent waves can be amplified to allow imaging further away from the sample. Research in this field is still early but very exciting.

Deep Learning and the Diffraction Limit?

Tantalizing hints suggest that deep learning may facilitate the spread of super-resolution microscopy. A few early papers have shown that it might be possible for deep learning algorithms to speed up the construction of super-resolution images or enable effective super-resolution with relatively cheap hardware. (We point to one such paper in the previous note.)

These hints are particularly compelling because deep learning can effectively perform tasks such as image deblurring.[2] This evidence suggests that it may be possible to build a robust set of super-resolution tools based on deep learning that could dramatically facilitate the adoption of such techniques. At present this research is still immature, and compelling tooling doesn't yet exist. However, we hope that this state of affairs will change over the coming years.

Preparing Biological Samples for Microscopy

One of the most critical steps in applying microscopy in the life sciences is preparing the sample for the microscope. This can be a highly nontrivial process that requires considerable experimental sophistication. We discuss a number of techniques for preparing samples in this section and comment on the ways in which such techniques can go wrong and create unexpected experimental artifacts.

Staining

The earliest optical microscopes allowed for magnified views of microscopic objects. This power enabled amazing improvements in the understanding of small objects, but it had the major limitation that it was not possible to highlight certain areas of the image for contrast. This led to the development of chemical stains which permitted scientists to view regions of the image for contrast.

A wide variety of stains have been developed to handle different types of samples. The staining procedures themselves can be quite involved, with multiple steps. Stains can be extraordinarily influential scientifically. In fact, it's common to classify to bacteria

---

2 Tao, Xin, et al. "Scale-Recurrent Network for Deep Image Deblurring." *https://arxiv.org/pdf/1802.01770.pdf*. 2018.

as "gram-positive" or "gram-negative" depending on their response to the well known Gram stain for bacteria. A task for a deep learning system might be to segment and label the gram-positive and gram-negative bacteria in microscopy samples. If you had a potential antibiotic in development, this would enable you to study its effect on gram-positive and gram-negative species separately.

Why Should I Care as a Developer?

Some of you reading this section may be developers interested in dealing with the challenges of building and deploying deep microscopy pipelines. You might reasonably be asking yourself whether you should care about the biology of sample preparation.

If you are indeed laser-focused on the challenges of building pipelines, skipping ahead to the case studies in this chapter will probably help you most. However, building an understanding of basic sample preparation may save you headaches later on and give you the vocabulary to effectively communicate with your biologist peers. If biology requests that you add metadata fields for stains, this section will give you a good idea of what they're actually asking for. That's worth a few minutes of your time!

Developing Antibacterial Agents for Gram-Negative Bacteria

One of the major challenges in drug discovery at this time is developing effective antibiotics for gram-negative bacteria. Gram-negative bacteria have an additional cell wall that prevents common antibacterial agents which target the peptidoglycan cell walls of gram-positive bacteria from functioning effectively.

This challenge is becoming more urgent because many bacterial strains are picking up gram-negative resistance through methods such as horizontal gene transfer, and deaths from bacterial infections are once again on the rise after decades of control.

It might well be possible to combine the deep learning methods for molecular design you've seen already with some of the imaging-based techniques you'll learn in this chapter to make progress on this problem. We encourage those of you curious about the possibilities to explore this area more carefully.

Sample Fixation

Large biological samples such as tissue will often degrade rapidly if left to their own devices. Metabolic processes in the sample will consume and damage the structure of the organs, cells, and organelles in the sample. The process of "fixation" seeks to stop this process, and stabilize the contents of the sample so that it can be imaged properly. A number of *fixative* agents have been designed which aid in this process. One of the core functions of fixatives is to denature proteins and turn off proteolytic enzymes. Such enzymes will consume the sample if allowed.

In addition, the process of fixation seeks to kill microorganisms that may damage the sample. For example, in heat fixation the sample is passed through a Bunsen burner. This process can damage the internal structures of the sample as a side effect. Another common technique is that of immersion fixation, where samples are immersed in a fixative solution and allowed to soak. For example, a sample could be soaked in cold formalin for a span of time, such as 24 hours.

Perfusion is a technique for fixing tissue samples from larger animals such as mice. Experimenters inject fixative into the heart and wait for the mouse to die before extracting the tissue sample. This process allows for the fixative agent to spread through the tissue naturally and often yields superior results.

Sectioning Samples

An important part of viewing a biological sample is being able to slice out a thin part of the sample for the microscope. There exist a number of ingenious tools to facilitate this process, including the microtome (see Figure 7-8), which slices biological samples into thin slices for easy viewing. The microtome has its limitations: it's hard to slice very small objects this way. For such small objects, it might be better to use a technique such as confocal microscopy instead.

It's worth pausing and asking why it's useful to know that devices such as a microtome exist. Well, let's say that as an engineer, you're constructing a pipeline to handle a number of brain imaging samples. The sample brain was likely sliced into thin pieces using a microtome or similar cutting device. Knowing the physical nature of this process will aid you if you're (for example) building a schema to organize such images consistently.

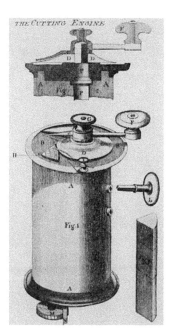

Figure 7-8. An early diagram from 1770 depicting a microtome. (Source: Wikimedia (https://commons.wikimedia.org/wiki/File:Cummings_1774_Microtome.jpg).)

Fluorescence Microscopy

A fluorescence microscope is an optical microscope that makes use of the phenomenon of fluorescence, where a sample of material absorbs light at one wavelength and emits it at another wavelength. This is a natural physical phenomenon; for example, a number of minerals fluoresce when exposed to ultraviolet light. It gets particularly interesting when applied to biology, though. A number of bacteria fluoresce naturally when their proteins absorb high-energy light and emit lower-energy light.

Fluorophores and fluorescent tags

A fluorophore is a chemical compound that can reemit light at a certain wavelength. These compounds are a critical tool in biology because they allow experimentalists to image particular components of a given cell in detail. Experimentally, the fluorophore is commonly applied as a dye to a particular cell. Figure 7-9 shows the molecular structure of a common fluorophore.

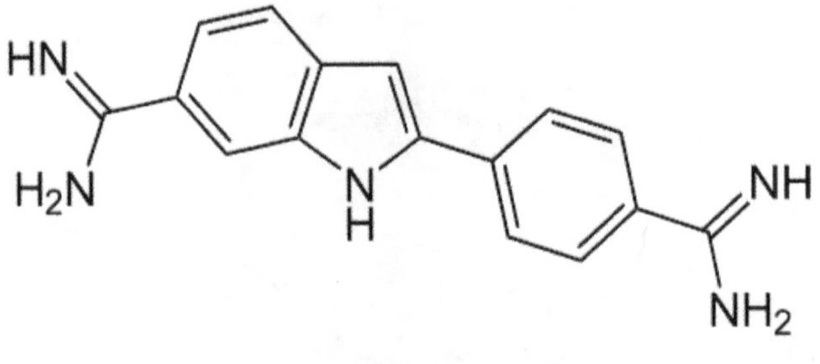

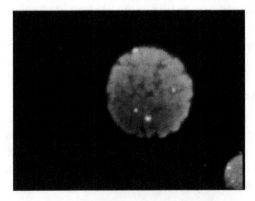

Figure 7-9. DAPI (4',6-diamidino-2-phenylindole) is a common fluorescent stain that binds to adenine-thymine–rich regions of DNA. Because it passes through cell membranes, it is commonly used to stain the insides of cells. (Source: Wikimedia (https:// commons.wikimedia.org/wiki/File:DAPI.svg).)

Fluorescent tagging is a technique for attaching a fluorophore to a biomolecule of interest in the body. There are a variety of techniques to do this effectively. It's useful in microscopy imaging, where it's common to want to highlight a particular part of the image. Fluorescent tagging can enable this very effectively.

Fluorescent microscopy has proven a tremendous boon for biological research because it permits researchers to zoom in on specific subsystems in a given biological sample, as opposed to dealing with the entirety of the sample. When studying individual cells, or individual molecules within a cell, the use of tagging can prove invaluable for focusing attention on interesting subsystems. Figure 7-10 shows how a fluorescent stain can be used to selectively visualize particular chromosomes within a human cell nucleus.

Figure 7-10. An image of a human lymphocyte nucleus with chromosomes 13 and 21 stained with DAPI (a popular fluorescent stain) to emit light. (Source: Wikimedia (https://commons.wikimedia.org/wiki/File:FISH_13_21.jpg).)

Fluorescence microscopy can be a very precise tool, used to track events like single binding events of molecules. For example, binding events of proteins to ligands (as discussed in Chapter 5) can be detected by a fluorescence assay.

Sample Preparation Artifacts

It's important to note that sample preparation can be a deeply tricky process. It's common for the preparation of the original sample to induce distortions in the object being imaged, which can lead to some confusion. An interesting example is the case of the mesosome, discussed in the following warning note.

The Mesosome: An Imaginary Organelle

The process of fixing a cell for electron microscopy introduces a crucial artifact, the *mesosome* in gram-positive bacteria (see Figure 7-11). Degradations in the cell wall, caused by the process of preparing the sample for the electron microscope, were originally thought to be natural structures instead of artifacts.

Be warned that similar artifacts likely exist in your own samples. In addition, it's entirely possible that a deep network could train itself to detect such artifacts rather than training itself to find real biology.

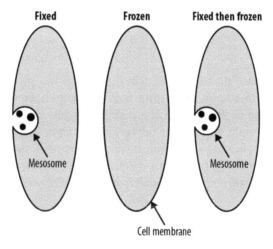

Figure 7-11. Mesosomes are artifacts introduced by preparation for electron microscopy that were once believed to be real structures in cells. (Adapted from Wikimedia (https:// en.wikipedia.org/wiki/Mesosome#/media/File:Mesosome_formation.svg).)

Tracking Provenance of Microscopy Samples

When you're designing systems to handle microscopy data, it will be critical to track the provenance of your samples. Each image should be annotated with information about the conditions in which it was gathered. This might include the physical device that was used to capture the image, the technician who supervised the imaging process, the sample that was imaged, and the physical location at which the sample was gathered. Biology is extraordinarily tricky to "debug." Issues such as the one described in the previous warning can go untracked, potentially for decades. Making sure to maintain adequate metadata around the provenance of your images could save you and your team from major issues down the line.

Deep Learning Applications

In this section we briefly review various applications of deep learning to microscopy, such as cell counting, cell segmentation, and computational assay construction. As we've noted previously in the chapter, this is a limited subset of the applications possible for deep microscopy. However, understanding these basic applications will give you the understanding needed to invent new deep microscopy applications of your own.

Cell Counting

A simple task is to count the number of cells that appear in a given image. You might reasonably ask why this is an interesting task. For a number of biological experiments, it can be quite useful to track the number of cells that survive after a given intervention. For example, perhaps the cells are drawn from a cancer cell line, and the intervention is the application of an anticancer compound. A successful intervention would reduce the number of living cancer cells, so it would be useful to have a deep learning system that can count the number of such living cells accurately without human intervention.

What Is a Cell Line?

Often in biology, it's useful to study cells of a given type. The first step in running an experiment against a collection of cells is to gain a plentiful supply of such cells. Enter the cell line. Cell lines are cells cultivated from a given source and that can grow stably in laboratory conditions.

Cell lines have been used in countless biological papers, but there are often serious concerns about the science done on them. To start, removing a cell from its natural

---

environment can radically change its biology. A growing line of evidence shows that the environment of a cell can fundamentally shape its response to stimuli.

Even more seriously, cell lines are often contaminated. Cells from one cell line may contaminate cells from another cell line, so results on a "breast cancer" cell line may, in fact, tell the researcher absolutely nothing about breast cancer!

For these reasons, studies on cell lines are often treated with caution, with the results intended simply as spur to attempt duplication with animal or human tests. Nevertheless, cell line studies provide an invaluable easy entry point to biological research and remain ubiquitous.

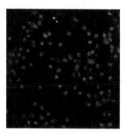

Figure 7-12. Samples of Drosophila cells. Note how the image conditions in microscopy images can be significantly different from image conditions in photographs. (Source: Cell Image Library (http://cellimagelibrary.org/images/21780).)

As Figure 7-12 shows, image conditions in cell microscopy can be significantly different from standard image conditions, so it is not immediately obvious that technologies such as convolutional neural networks can be adapted to tasks such as cell counting. Luckily, significant experimental work has shown that convolutional networks do quite well at learning from microscopy datasets.

Implementing cell counting in DeepChem

This section walks through the construction of a deep learning model for cell counting using DeepChem. We'll start by loading and featurizing a cell counting dataset. We use the Broad Bioimage Benchmark Collection (*https://data.broadinstitute.org/bbbc/*) (BBBC) to get access to a useful microscopy dataset for this purpose.

BBBC Datasets

The BBBC datasets contain a useful collection of annotated bioimage datasets from various cellular assays. It is a useful resource as you work on training your own deep microscopy models. DeepChem has a collection of image processing resources to make it easier to work with these datasets. In particular, DeepChem's Image Loader class facilitates loading of the datasets.

Processing Image Datasets

Images are usually stored on disk in standard image file formats (PNG, JPEG, etc.). The processing pipeline for image datasets typically reads in these files from disk and transforms them into a suitable in-memory representation, typically a multidimensional array. In Python processing pipelines, this array is often simply a NumPy array. For an image N pixels high, M pixels wide, and with 3 RGB color channels, you would get an array of shape $(N, M, 3)$. If you have 10 such images, these images would typically be batched into an array of shape $(10, N, M, 3)$.

Before you can load the dataset into DeepChem, you'll first need to download it locally. The BBBC005 dataset that we'll use for this task is reasonably large (a little under 2 GB), so make sure your development machine has sufficient space available:

```
wget https://data.broadinstitute.org/bbbc/BBBC005/BBBC005_v1_images.zip
unzip BBBC005_v1_images.zip
```

With the images downloaded to your local machine, you can now load them into DeepChem by using `ImageDataset`:

```
image_dir = 'BBBC005_v1_images'
files = []
labels = []
for f in os.listdir(image_dir):
  if f.endswith('.TIF'):
    files.append(os.path.join(image_dir, f))
    labels.append(int(re.findall('_C(.*?)_', f)[0]))
dataset = dc.data.ImageDataset(files, np.array(labels))
```

This code walks through the downloaded directory of images and pulls out the image files. The labels are encoded in the filenames themselves, so we use a simple regular expression to extract the number of cells in each image. We use `ImageDataset` to load them as a DeepChem dataset.

Let's now split this dataset into training, validation, and test sets:

```
splitter = dc.splits.RandomSplitter()
train_dataset, valid_dataset, test_dataset = splitter.train_valid_test_split(
        dataset, seed=123)
```

With this split in place, we can now define the model itself. In this case, let's make a simple convolutional architecture with a fully connected layer at the end:

```
features = tf.keras.Input(shape=(520, 696, 1))
prev_layer = features
for num_outputs in [16, 32, 64, 128, 256]:
  prev_layer = layers.Conv2D(num_outputs, kernel_size=5, strides=2,
                             activation=tf.nn.relu)(prev_layer)
output = layers.Dense(1)(layers.Flatten()(prev_layer))
```

```
keras_model = tf.keras.Model(inputs=features, outputs=output)
learning_rate = dc.models.optimizers.ExponentialDecay(0.001, 0.9, 250)
model = dc.models.KerasModel(
    keras_model,
    loss=dc.models.losses.L2Loss(),
    learning_rate=learning_rate,
    model_dir='model')
```

Note that we use L2Loss to train our model as a regression task. Even though cell counts are whole numbers, we don't have a natural upper bound on the number of cells in an image.

Training this model will take some computational effort (more on this momentarily), so to start, we recommend using our pretrained model for basic experimentation. This model can be used to make predictions out of the box. There are directions on downloading the pretrained model in the code repository associated with the book (*https://github.com/deepchem/DeepLearningLifeSciences*). Once you've downloaded it, you can load the pretrained weights into the model with:

```
model.restore()
```

Let's take this pretrained model out for a whirl. First, we'll compute the average prediction error on our test set for our cell counting task:

```
y_pred = model.predict(test_dataset).flatten()
print(np.sqrt(np.mean((y_pred-test_dataset.y)**2)))
```

What accuracy do you get when you try running the model?

Now, how can you train this model for yourself? You can fit the model by training it for 50 epochs on the dataset:

```
model.fit(train_dataset, nb_epoch=50)
```

This learning task will take some amount of computing horsepower. On a good GPU, it should complete within an hour or so. It may not be feasible to easily train the model on a CPU system.

Once trained, test the accuracy of the model on the validation and test sets. Does it match that of the pretrained model?

Cell Segmentation

The task of cellular segmentation involves annotating a given cellular microscopy image to denote where cells appear and where background appears. Why is this useful? If you recall our earlier discussion of gram-positive and gram-negative bacteria, you can probably guess why an automated system for separating out the two types of bacteria might prove useful. It turns out that similar problems arise through all of cellular microscopy (and in other fields of imaging, as we will see in Chapter 8).

Segmentation masks provide significantly finer-grained resolution and permit for more refined analysis than cell counting. For example, it might be useful to understand what fraction of a given plate is covered with cells. Such analysis is easy to perform once segmentation masks have been generated. Figure 7-13 provides an example of a segmentation mask that is generated from a synthetic dataset.

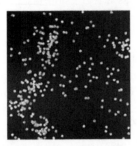 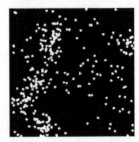

Figure 7-13. A synthetic dataset of cells (on the left) along with foreground/background masks annotating where cells appear in the image. (Source: Broad Institute (https:// data.broadinstitute.org/bbbc/BBBC005/).)

That said, segmentation asks for significantly more from a machine learning model than counting. Being able to precisely differentiate cellular and noncellular regions requires greater precision in learning. For that reason, it's not surprising that machine learning segmentation approaches are still harder to get working than simpler cellular counting approaches. We will experiment with a segmentation model later in this chapter.

Where Do Segmentation Masks Come From?

It's worth pausing to note that segmentation masks are complex objects. There don't exist good algorithms (except for deep learning techniques) for generating such masks in general. How then can we bootstrap the training data needed to refine a deep segmentation technique? One possibility is to use synthetic data, as in Figure 7-13. Because the cellular image is generated in a synthetic fashion, the mask can also be synthetically generated. This is a useful trick, but it has obvious limitations because it will limit our learned segmentation methods to similar images.

A more general procedure is to have human annotators generate suitable segmentation masks. Similar procedures are used widely to train self-driving cars. In that task, finding segmentations that annotate pedestrians and street signs is critical, and armies of human segmenters are used to generate needed training data. As machine-learned microscopy grows in importance, it is likely that similar human pipelines will become critical.

Implementing cell segmentation in DeepChem

In this section, we will train a cellular segmentation model on the same BBBC005 dataset that we used previously for the cell counting task. There's a crucial subtlety here, though. In the cell counting challenge, each training image has a simple count as a label. However, in the cellular segmentation task, each label is itself an image. This means that a cellular segmentation model is actually a form of "image transformer" rather than a simple classification or regression model. Let's start by obtaining this dataset. We have to retrieve the segmentation masks from the BBBC website, using the following commands:

```
wget https://data.broadinstitute.org/bbbc/BBBC005/BBBC005_v1_ground_truth.zip
unzip BBBC005_v1_ground_truth.zip
```

The ground-truth data is something like 10 MB, so it should be easier to download than the full BBBC005 dataset. Now let's load this dataset into DeepChem. Luckily for us, ImageDataset is set up to handle image segmentation datasets without much extra hassle:

```
image_dir = 'BBBC005_v1_images'
label_dir = 'BBBC005_v1_ground_truth'
rows = ('A', 'B', 'C', 'D', 'E', 'F', 'G', 'H', 'I', 'J', 'K', 'L',
        'M', 'N', 'O', 'P')
blurs = (1, 4, 7, 10, 14, 17, 20, 23, 26, 29, 32, 35, 39, 42, 45, 48)
files = []
labels = []
for f in os.listdir(label_dir):
  if f.endswith('.TIF'):
    for row, blur in zip(rows, blurs):
```

```
        fname = f.replace('_F1', '_F%d'%blur).replace('_A', '_%s'%row)
        files.append(os.path.join(image_dir, fname))
        labels.append(os.path.join(label_dir, f))
    dataset = dc.data.ImageDataset(files, labels)
```

Now that we have our datasets loaded and processed, let's hop into building some
deep learning models for them. As before, we'll split this dataset into training, validation, and test sets:

```
splitter = dc.splits.RandomSplitter()
train_dataset, valid_dataset, test_dataset = splitter.train_valid_test_split(
        dataset, seed=123)
```

What architecture can we use for the task of image segmentation? It's not just a matter of using a straightforward convolutional architecture since our output needs to
itself be an image (the segmentation mask). Luckily for us, there has been some past
work on suitable architectures for this task. The U-Net architecture uses a stacked
series of convolutions to progressively "downsample" and then "upsample" the source
image, as illustrated in Figure 7-14. This architecture does well at the task of image
segmentation.

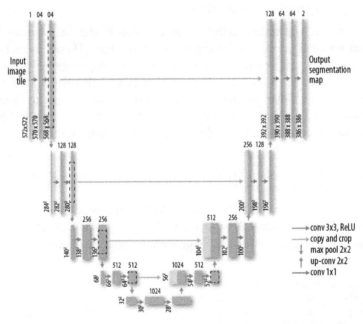

Figure 7-14. A representation of the U-Net architecture for biomedical image segmenta-
tion. (Adapted from the University of Freiburg (https://lmb.informatik.uni-freiburg.de/
people/ronneber/u-net/).)

Let's now implement the U-Net in DeepChem:

```
features = tf.keras.Input(shape=(520, 696, 1))
# Downsample three times.
conv1 = layers.Conv2D(16, kernel_size=5, strides=2,
            activation=tf.nn.relu, padding='same')(features/255.0)
conv2 = layers.Conv2D(32, kernel_size=5, strides=2,
            activation=tf.nn.relu, padding='same')(conv1)
conv3 = layers.Conv2D(64, kernel_size=5, strides=2,
            activation=tf.nn.relu, padding='same')(conv2)
# Do a 1x1 convolution.
conv4 = layers.Conv2D(64, kernel_size=1, strides=1)(conv3)
# Upsample three times.
concat1 = layers.Concatenate(axis=3)([conv3, conv4])
deconv1 = layers.Conv2DTranspose(32, kernel_size=5, strides=2,
            activation=tf.nn.relu, padding='same')(concat1)
concat2 = layers.Concatenate(axis=3)([conv2, deconv1])
deconv2 = layers.Conv2DTranspose(16, kernel_size=5, strides=2,
            activation=tf.nn.relu, padding='same')(concat2)
concat3 = layers.Concatenate(axis=3)([conv1, deconv2])
deconv3 = layers.Conv2DTranspose(1, kernel_size=5, strides=2,
            activation=tf.nn.relu, padding='same')(concat3)
# Compute the final output.
concat4 = layers.Concatenate(axis=3)([features, deconv3])
logits = layers.Conv2D(1, kernel_size=5, strides=1, padding='same')(concat4)
output = layers.Activation(tf.math.sigmoid)(logits)
keras_model = tf.keras.Model(inputs=features, outputs=[output, logits])
learning_rate = dc.models.optimizers.ExponentialDecay(0.01, 0.9, 250)
model = dc.models.KerasModel(
    keras_model,
    loss=dc.models.losses.SigmoidCrossEntropy(),
    output_types=['prediction', 'loss'],
    learning_rate=learning_rate,
    model_dir='models/segmentation')
```

This architecture is somewhat more complex than that for cell counting, but we use the same basic code structure and stack convolutional layers to achieve our desired architecture. As before, let's use a pretrained model to give this architecture a try. Directions for downloading the pretrained model are available in the book's code repository (*https://github.com/deepchem/DeepLearningLifeSciences*). Once you've got the pretrained weights in place, you can load the weights as before:

```
model.restore()
```

Let's use this model to create some masks. Calling `model.predict_on_batch()` allows us to predict the output mask for a batch of inputs. We can check the accuracy of our predictions by comparing our masks against the ground-truth masks and checking the overlap fraction:

```
scores = []
for x, y, w, id in test_dataset.itersamples():
  y_pred = model.predict_on_batch([x]).squeeze()
```

```
    scores.append(np.mean((y>0) == (y_pred>0.5)))
  print(np.mean(scores))
```

This should return approximately 0.9899. This means nearly 99% of pixels are correctly predicted! It's a neat result, but we should emphasize that this is a toy learning task. A simple image processing algorithm with a brightness threshold could likely do almost as well. Still, the principles exposed here should carry over to more complex image datasets.

OK, now that we've explored with the pretrained model, let's train a U-Net from scratch for 50 epochs and see what results we obtain:

```
  model.fit(train_dataset, nb_epoch=50, checkpoint_interval=100)
```

As before, this training is computationally intensive and will take a couple of hours on a good GPU. It may not be feasible to train this model on a CPU system. Once the model is trained, try running the results for yourself and seeing what you get. Can you match the accuracy of the pretrained model?

Computational Assays

Cell counting and segmentation are fairly straightforward visual tasks, so it's perhaps unsurprising that machine learning models are capable of performing well on such datasets. It could reasonably be asked if that's all machine learning models are capable of.

Luckily, it turns out the answer is no! Machine learning models are capable of picking up on subtle signals in the dataset. For example, one study demonstrated that deep learning models are capable of predicting the outputs of fluorescent labels from the raw image.[3] It's worth pausing to consider how surprising this result is. As we saw in "Preparing Biological Samples for Microscopy" on page 110, fluorescent staining can be a considerably involved procedure. It's astonishing that deep learning might be able to remove some of the needed preparation work.

This is an exciting result, but it's worth noting that it's still an early one. Considerable work will have to be done to "robustify" these techniques so they can be applied more broadly.

Conclusion

In this chapter, we've introduced you to the basics of microscopy and to some basic machine learning approaches to microscopy systems. We've provided a broad introduction to some of the fundamental questions of modern microscopy (especially as

---

3 Christensen, Eric. "In Silico Labeling: Predicting Fluorescent Labels in Unlabeled Images." *https://github.com/ google/in-silico-labeling.*

applied to biological problems) and have discussed where deep learning has already had an impact and hinted at places where it could have an even greater impact in the future.

We've also provided a thorough overview of some of the physics and biology surrounding microscopy, and tried to convey why this information might be useful even if to developers who are primarily interested in building effective pipelines for handling microscopy images and models. Knowledge of physical principles such as the diffraction limit will allow you to understand why different microscopy techniques are used and how deep learning might prove critical for the future of the field. Knowledge of biological sample preparation techniques will help you understand the types of metadata and annotations that will be important to track when designing a practical microscopy system.

While we're very excited about the potential applications of deep learning techniques in microscopy, it's important for us to emphasize that these methods come with a number of caveats. For one, a number of recent studies have highlighted the brittleness of visual convolutional models.[4] Simple artifacts can trip up such models and cause significant issues. For example, the image of a stop sign could be slightly perturbed so that a model classifies it as a green traffic light. That would be disastrous for a self-driving car!

Given this evidence, it's worth asking what the potential pitfalls are with models for deep microscopy. Is it possible that deep microscopy models are simply backfilling from memorized previous data points? Even if this isn't the entire explanation for their performance, it is likely that at least part of the power of such deep models comes from such regurgitation of memorized data. This could very well result in imputation of spurious correlations. As a result, when doing scientific analysis on microscopic datasets, it will be critical to stop and question whether your results are due to model artifacts or genuine biological phenomena. We will provide some tools for you to critically probe models in upcoming chapters so you can better determine what your model has actually learned.

In the next chapter, we will explore applications of deep learning for medicine. We will reuse many of the skills for visual deep learning that we covered in this chapter.

---

4 Rosenfeld, Amir, Richard Zemel, and John K. Tsotsos. "The Elephant in the Room." *https://arxiv.org/abs/1808.03305*. 2018.

## CHAPTER 8

Deep Learning for Medicine

As we saw in the previous chapter, the ability to extract meaningful information from visual datasets can prove useful for analyzing microscopy images. This capability for handling visual data is similarly useful for medical applications. Much of modern medicine requires doctors to critically analyze medical scans. Deep learning tools could potentially make this analysis easier and faster (but perhaps less interpretable).

Let's learn more. We'll start by giving you a brief overview of earlier computational techniques for medicine. We'll discuss some of the limitations of such methods, then we'll start running over the current set of deep learning–powered techniques for medicine. We'll explain how these new techniques might allow us to bypass some of the fundamental limitations of older techniques. We'll end the chapter with a discussion of some of the ethical considerations of applying deep learning to medicine.

Computer-Aided Diagnostics

Designing computer-aided diagnostic systems has been a major focus of AI research since the advent of the field. The earliest attempts at this[1] used hand-curated knowledge bases. In these systems, expert doctors would be solicited to write down causal inference rules (see, for example, Figure 8-1).

There was basic support for uncertainty handling through certainty factors.

---

1 See Dendral (*https://en.wikipedia.org/wiki/Dendral*) or Mycin (*https://en.wikipedia.org/wiki/Mycin*) on Wikipedia for more information.

IF
1) **The stain of the organism in grampos, and** (01)
2) **The morphology of the organism is COCCUS, and** (02)
3) **The growth confirmation of the organisms chains** (03)

THEN
There is suggestive evidence (0.7) that the identity
of the organism is streptococus. (h1)

Figure 8-1. MYCIN was an early expert system used to diagnose bacterial infections.
This is an example of a MYCIN rule for inference (adapted from the University of Surrey
(http://www.computing.surrey.ac.uk/ai/PROFILE/mycin.html#Certainity%20Factors)).

These rules were combined using a logical engine. A number of efficient inference techniques were designed that could effectively combine large databases of rules. Such systems were traditionally called "expert systems."

What Happened to Expert Systems?

Although expert systems achieved some notable successes, the construction of these systems required considerable effort. Rules had to be painstakingly solicited from experts and curated by trained "knowledge engineers." While some expert systems achieved striking results in limited domains, on the whole they were too brittle to use widely. That said, expert systems had a strong impact on much of computer science, and hosts of modern technologies (SQL, XML, Bayesian networks and more) draw inspiration from expert system technologies.

If you're a developer, it's good to pause and consider this. Although expert systems were once a blindingly hot technology, they currently exist primarily as a historical curiosity. It's very likely that most of today's hot technologies will one day end up in the curiosity heap of computer science history. This is a feature, not a bug, of computer science. The field reinvents itself rapidly, so we can trust that the replacements for today's technologies will tick some crucial boxes that today's tools can't. At the same time, as with expert systems, we can rest assured that the algorithmic fundamentals of today's technology will live on in tomorrow's tools.

Expert systems for medicine had a good run. Some of them were deployed widely and adopted internationally as well.[2] However, these systems failed to achieve significant traction with everyday doctors and nurses. One problem was that they were very finicky and hard to use. They also required their users to be able to pass in patient information in a highly structured format. Given that computers had barely penetrated standard clinics at the time, requiring highly specialized training for doctors and nurses proved to be too big an ask.

Probabilistic Diagnoses with Bayesian Networks

Another major problem with expert system tools was that they could only provide deterministic predictions. These deterministic predictions didn't leave much room for uncertainty. What if the doctor was seeing a tricky patient where the diagnosis wasn't clear? For a time, it seemed that if expert systems could be modified to account for uncertainties, this would allow them to achieve success.

This basic insight triggered a host of work on Bayesian networks for clinical diagnoses. (One of the authors of this book spent a year working on such a system as an undergrad.) However, these systems suffered from many of the same limitations as the expert systems. It was still necessary to solicit structural knowledge from doctors, and designers of Bayesian clinical networks faced the additional challenge of soliciting meaningful probabilities from doctors. This process added significant overhead to the process of adoption.

In addition, training a Bayesian network can be complicated. Different types of Bayesian network require different algorithms. Contrast this with deep learning algorithms, where gradient descent techniques work on almost all networks you can find. Robustness of learning is often what enables widespread adoption. This basic insight triggered a host of work on Bayesian networks for clinical diagnoses. (See Figure 8-2 for a simple example of a Bayesian network.)

2 Asabere, Nana Yaw. "mMes: A Mobile Medical Expert System for Health Institutions in Ghana." *International Journal of Science and Technology* no.6. (June 2012). *https://pdfs.semanticscholar.org/ed35/ ec162c5916f317162e11e390440bdb1b55b2.pdf.*

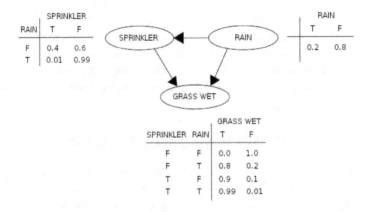

RAIN	SPRINKLER T	F
F	0.4	0.6
T	0.01	0.99

RAIN T	F
0.2	0.8

SPRINKLER	RAIN	GRASS WET T	F
F	F	0.0	1.0
F	T	0.8	0.2
T	F	0.9	0.1
T	T	0.99	0.01

Figure 8-2. A simple example of a Bayesian network for inferring whether the grass is wet at a given spot. (Source: Wikimedia (https://commons.wikimedia.org/wiki/ File:SimpleBayesNet.svg).)

Ease of Use Drives Adoption

Expert systems and Bayesian networks both failed to win broad adoption. At least part of the reason for this failure was that both these systems had pretty terrible developer experiences. From the developer's standpoint, designing either a Bayesian network or an expert system required constantly keeping a doctor in the development loop. In addition, the effectiveness of the system depended critically on the ability of the development team to extract valuable insights from doctors.

Contrast this with deep networks. For a given data type (images, molecules, text, etc.) and a given learning task, there are a set of standard metrics at hand. The developer needs only to follow best statistical practices (as taught by this or another book) in order to build a functional system. The dependence on expert knowledge is considerably reduced. This gain in simplicity no doubt accounts for part of the reason deep networks have gained much broader adoption.

Electronic Health Record Data

Traditionally, hospitals maintained paper charts for their patients. These charts would record the tests, medications, and other treatments of the patient, allowing doctors to track the patient's health with a quick glance at the chart. Unfortunately, paper health records had a host of difficulties associated with them. Transferring records between hospitals required a major amount of work, and it wasn't easy to index or search paper health record data.

For this reason, there has been a major push over the last few decades in a number of countries to move from paper records to electronic health records (EHRs). In the US, the adoption of the Affordable Care Act significantly accelerated their adoption, and most major US health providers now store their patient records on EHR systems.

The broad adoption of EHR systems has spurred a boom in research on machine learning systems that work with EHR data. These systems aim to use large datasets of patient records to train models that will be capable of predicting things such as patient outcomes or risks. In many ways, these EHR models are the intellectual successors of the expert systems and Bayesian networks we just learned about. Like these earlier systems, EHR models seek to aid the process of diagnosis. However, while earlier systems sought to aid doctors in making real-time diagnoses, these newer systems content themselves (mostly) with working on the backend.

A number of projects have attempted to learn robust models from EHR data. While there have been some notable successes, learning on EHR data remains challenging for practitioners. Due to privacy concerns, there aren't many large public EHR datasets available. As a result, only a small group of elite researchers have been able to design these systems thus far. In addition, EHR data tends to be very messy. Since human doctors and nurses manually enter information, most EHR data suffers from missing fields and all sorts of different conventions. Creating robust models that deal with the missing data has proven challenging.

ICD-10 Codes

ICD-10 is a set of "codes" for patient diseases and symptoms. These standard codes have found broad adoption in recent years because they allow insurers and governmental agencies to set standard practices, treatments, and treatment prices for diseases.

The ICD-10 codes "quantize" (make discrete) the high-dimensional continuous space of human disease. By standardizing, they allow doctors to compare and group patients. It's worth noting that for this reason, such codes will likely prove relevant to developers of EHR systems and models. If you're designing the data warehouse for a new EHR system, make sure you think about where you're going to put your codes!

Fast Healthcare Interoperability Resources (FHIR)

The Fast Healthcare Interoperability Resources (FHIR) specification was developed to represent clinical data in a standard and flexible format.[3] Recent work from Google demonstrated how raw EHR data can be transformed into FHIR format automatically.[4] The use of this format enables the development of standard deep architectures that can be applied to arbitrary EHR data, which means standard open source tools for this data can be used in a plug-and-play fashion. This work is still in early stages, but it represents exciting progress for the field. Although standardization may appear boring at first blush, it's the foundation for future advances since it means that larger datasets can be worked with productively.

However, this state of affairs is starting to change. Improved tools, both for preprocessing and for learning, have started to enable effective learning to occur on EHR systems. The DeepPatient system trains a denoising autoencoder on patient medical records to create a patient representation which it then uses to predict patient outcomes.[5] In this system, a patient's record is transformed from a set of unordered textual information into a vector. This strategy of transforming disparate data types into vectors has been widely successful throughout deep learning and seems poised to offer meaningful improvements in EHR systems as well. A number of models based on EHR systems have sprouted in the literature, many of which are starting to incorporate the latest tools of deep learning, such as recurrent networks or reinforcement learning. While models with these latest bells and whistles are still maturing, they're very exciting and provide pointers to where the field is likely to head over the next few years.

What About Unsupervised Learning?

Through most of this book, we've primarily demonstrated supervised learning methods. There's also a whole class of "unsupervised" learning methods that don't share the same dependency on supervised training data. We haven't really introduced unsupervised learning as a concept yet, but the basic idea is that we no longer have labels

---

3 Mandel, JC, et al. "SMART on FHIR: A Standards-Based, Interoperable Apps Platform for Electronic Health Records." *https://doi.org/10.1093/jamia/ocv189*. 2016.

4 Rajkomar, Alvin et al. "Scalable and Accurate Deep Learning with Electronic Health Records." *NPJ Digital Medicine. https://arxiv.org/pdf/1801.07860.pdf*. 2018.

5 Miotto, Riccardo, Li Li, Brian A. Kidd and Joel T. Dudley. "Deep Patient: An Unsupervised Representation to Predict the Future of Patients from the Electronic Health Records." *https://doi.org/10.1038/srep26094*. 2016.

associated with data points. For example, imagine we have a set of EHR records but no patient outcome data. What can we do?

The simplest answer is that we can *cluster* the records. For a toy example, imagine we have "twin" patients whose EHR records are identical. It seems reasonable to predict that the outcomes of these two patients will be similar. Unsupervised learning techniques such as *k*-means or autoencoders implement somewhat more sophisticated forms of this basic intuition. You'll see a sophisticated example of an unsupervised algorithm later in Chapter 9.

Unsupervised techniques can yield some compelling insights, but these methods can be hit-or-miss at times. While there have been some compelling use cases, such as DeepPatient, on the whole unsupervised methods are still finicky enough that they have yet to see wide usage. If you're a researcher, though, working on ways to stabilize unsupervised learning remains a compelling (and challenging) open problem.

The Dangers of Large Patient EHR Databases?

A number of large institutions are moving toward having all their patients in EHR systems. What happens when these large datasets are standardized (perhaps in a format such as FHIR) and made interoperable? On the positive side, it might be possible then to support applications such as searching for patients that have a particular disease phenotype. Such focused search capabilities may help doctors find treatments more effectively for patients, and especially for patients with rare diseases.

However, it doesn't take much imagination to see how large patient databases could be put to malicious use. For example, insurers could use patient outcome systems to preemptively deny insurance to higher-risk patients, or top surgeons seeking to maintain high patient survival rates could avoid operating on patients that the system marks as high risk. How do we guard against these dangers?

Many of the questions that machine learning systems raise can't be addressed with the tools of machine learning. Rather, it's likely that the answers to these questions will rest with legislation that forbids predatory behavior on the part of doctors, insurers, and others.

Do EHRs Really Help Doctors?

While EHRs obviously aid in the design of learning algorithms, there's less compelling evidence that EHRs actually improve life for doctors. Part of the challenge is that today's EHRs require significant manual data entry on the part of doctors. For patients, this has created a new familiar dynamic in which the doctor spends the majority of a consultation looking at the computer rather than looking at the actual patient.

This state of affairs has left both patients and doctors unhappy.[6] Doctors feel burned out because they spend the majority of their time doing clerical data entry rather than patient care, and patients feel ignored. One hope for the next generation of deep learning–powered systems is that this imbalance could be improved by future products.

Note, however, that there's a real chance that the next generation of deep learning tools could prove equally unfriendly and unhelpful for doctors. The designers of EHR systems didn't aim to make unfriendly systems either.

Deep Radiology

Radiology is the science of using medical scans to diagnose disease. There are a variety of different scans that doctors use, such as MRI scans, ultrasounds, X-rays, and CT scans. For each of these, the challenge is to diagnose the state of the patient from the given scan imagery. This looks like a challenge well suited for convolutional learning methods. As we have seen in the previous chapters, deep learning methods are capable of learning sophisticated functions from image data. Much of modern radiology (the mechanical parts at least) consists of classifying and handling complex medical image data. The use of scans has a long and storied history in medicine (see Figure 8-4 for an example of an early X-ray).

In this section, we'll quickly introduce a number of different types of scans and briefly cover some deep learning applications. Many of these applications are qualitatively similar. They start by obtaining a large enough dataset of scans from a medical institution. These scans are used to train a convolutional architecture (see Figure 8-3). Often, the architecture is a standard VGG or ResNet architecture, but sometimes with some tweaks to the core structure. The trained model often (at least according to perhaps naive statistics) has strong performance on the task in question.

---

6 Gawande, Atul. "Why Doctors Hate Their Computers." *The New Yorker. https://www.newyorker.com/maga-zine/2018/11/12/why-doctors-hate-their-computers.* 2018.

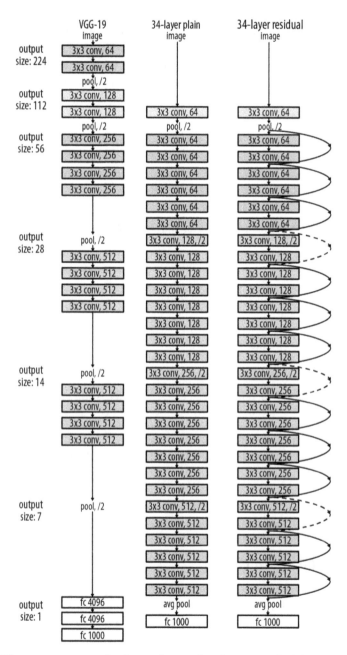

Figure 8-3. Here are some standard convolutional architectures (VGG-19, Resnet-34; number indicates the number of convolutions applied). These architectures are standard for image tasks and are commonly used for medical applications.

These advances have led to some perhaps inflated expectations. Some high-profile AI scientists—most notably, Geoff Hinton—have commented that deep learning for radiology will advance so far that it will no longer be worth training new radiologists in the near future.[7] Is this actually right? There have been a string of recent advances in which deep learning systems have achieved what appears like near-human performance. However, these results come with many caveats, and these systems are often brittle in unknown fashions.

Our opinion is that the risk of direct 1-1 replacement of doctors remains low, but there is a real risk of systematic displacement. What does this mean? New startups are working to invent new business models in which deep learning systems do the large majority of scan analysis, with only a few doctors remaining.

 Is Deep Learning Actually Learning Medicine?

Significant analysis has gone into scrutinizing what deep models actually learn in medical imagery. Unfortunately, in many cases, it looks like the deep models succeed in picking up nonmedical factors in the imagery. For example, the model might implicitly learn to identify the scanning center in which a particular medical scan was taken. Since particular centers are often used for more serious or less serious patients, the model might at first glance look as though it had succeeded in learning useful medicine, but would in fact be generally useless.

What can be done in such cases? The jury is still out on the question, but a couple of early approaches are emerging. The first is to use the growing literature on model interpretability to scrutinize carefully what the model is learning. In Chapter 10, we will delve into a number of methods for model interpretability.

The other approach is to conduct prospective trials deploying the models in clinics. Prospective trials remain the gold standard for testing proposed medical interventions, and it is likely they will remain so for deep learning techniques as well.

X-Ray Scans and CT Scans

Informally, an X-ray scan—radiography, if we're being precise—involves using X-rays to view some internal structure in the body (Figure 8-4). Computed tomography (CT) scans are a variant of X-ray scans in which the X-ray source and detectors rotate around the object being imaged, allowing for 3D images.

---

7 "AI, Radiology and the Future of Work." *The Economist. https://econ.st/2HrRDuz.* 2018.

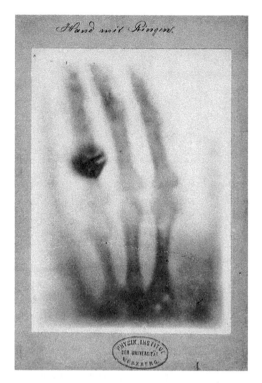

Figure 8-4. The first medical X-ray taken by Wilhelm Röntgen of his wife Anna Bertha Ludwig's hand. The science of X-rays has come a long way since this first photograph, and there's a chance that deep learning will take it much further yet!

A common misconception is that X-ray scans are only capable of imaging "hard" objects such as bones. This turns out to be quite false. CT scans are routinely used to image tissues in the body such as the brain (Figure 8-5), and backscatter X-rays are often used in airports to image travelers at security checkpoints. Mammograms use low-energy X-rays to scan breast tissue as well.

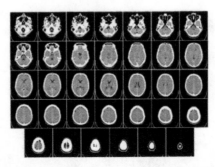

Figure 8-5. CT scan of a human brain from bottom to top. Note the capacity of CT scans to provide 3D information. (Source: Wikimedia (https://commons.wikimedia.org/wiki/File:Computed_tomography_of_human_brain_-_large.png).)

It's worth noting that all X-ray scans are known to be linked to cancer, so a common goal is to minimize the exposure of patients to radiation by limiting the number of scans required. This risk is more marked for CT scans, which have to expose the patient for longer time periods in order to gather sufficient data. A wide variety of signal processing algorithms have been designed to reduce the number of scans required for CT. Some exciting recent work has started to use deep learning to further tune this reconstruction process so even fewer scans are required.

However, most uses of deep learning in the space are used to classify scans. For example, convolutional networks have been used to classify Alzheimer progression from CT brain images.[8] Other work has claimed the ability to diagnose pneumonia from chest X-ray scans at near physician-level accuracy.[9] Deep learning has similarly been used to achieve strong classification accuracy on mammography.[10]

---

8 Gao, Xiaohong W., Rui Hui, and Zengmin Tian. "Classification of CT Brain Images Based on Deep Learning Networks." *https://doi.org/10.1016/j.cmpb.2016.10.007.* 2017.

9 Pranav Rajpurkar et al. "CheXNet: Radiologist-Level Pneumonia Detection on Chest X-Rays with Deep Learning." *https://arxiv.org/pdf/1711.05225.pdf.* 2017.

10 Ribli, Dezso et al. "Detecting and Classifying Lesions in Mammograms with Deep Learning." *https://doi.org/10.1038/s41598-018-22437-z.* 2018.

Human-Level Accuracy Is Tricky!

When a paper claims that its system achieves near-human accuracy, it's worth pausing to consider what that means. Usually, the authors of the paper choose some metric (say, ROC AUC), and a group of external physicians works to annotate the chosen test set for the study. The accuracy of the model on this test set is then compared against that of the "average" physician (often the mean or median of the physician scores).

This is a fairly complex process, and there are a number of ways in which this comparison can go wrong. First, the choice of metric can play a role—all too commonly, varying the choice of metric can lead to differences. Good analyses will consider multiple different metrics to ensure that the conclusions are robust to such variance.

Another point to note is that there's considerable variation between doctors themselves. It's worth checking to make sure that your choice of "average" is robust. A better metric might be to ask whether your algorithm is capable of beating the "best" doctor in the panel.

A third issue is that it can be extremely tricky to make sure that the test set isn't "polluted." (See our warning in Chapter 7.) Subtle forms of pollution can occur in which the scans from the same patient accidentally end up in both the training and test sets. If your model has very high accuracy, it's worth double- and triple-checking for such leakages. All of us have been guilty of making mistakes on these pipelines in the past.

Finally, "human-level accuracy" doesn't often mean much. As we've noted, some expert systems and Bayesian networks achieved human-level accuracy on limited tasks but failed to have a broad impact on medicine. The reason is that doctors perform a whole range of tasks which are tightly wound together. A given doctor may underperform a deep network on scan reading, but may be able to offer a much better diagnosis using other information. It's worth remembering that these tasks are often synthetic, and may not match best physician practices. Prospective clinical trials using deep learning systems live with consenting patients will be needed to more accurately gauge the effectiveness of these techniques.

Histology

Histology is the study of tissues, often viewed through microscopic scans. We won't say too much about it, because the issues that confront designers of deep histology systems are a subset of the issues that deep microscopy faces. Take a look back at that chapter to learn more. We'll note simply that deep learning models have achieved strong performance on histology studies.

MRI Scans

Magnetic resonance imaging (MRI) is another form of scan commonly used by doctors. Instead of X-rays, it uses strong magnetic fields to do its imaging. One advantage of MRI scans is therefore limited radiation exposure. However, these scans often require patients to lie within a noisy and cramped MRI machine, an experience which may be considerably more unpleasant for patients than X-ray scans.

Like CT, MRI is capable of assembling 3D images. And as with CT scans, a number of deep learning studies have sought to ease this reconstruction process. Some early studies claim that deep learning techniques can improve on traditional signal processing methods to reconstruct MRI images with reduced scan times. In addition, as with other scanning techniques, a number of studies have sought to use deep networks for classifying, segmenting, and processing MRI images with some strong successes.

Deep Learning for Signal Processing?

For both CT scans and MRI scans, we've mentioned in passing that deep networks have been used to help reconstruct images more effectively. Both of these applications are examples of the broader trend of using deep learning in signal processing. We've already seen some of this in passing; deep learning methods for super-resolution microscopy also fall within this general framework.

Such work on improving signal processing techniques is very exciting from a fundamental perspective, since signal processing is a highly mathematical, developed field. The fact that deep learning offers new directions here is in itself groundbreaking! However, it's also worth noting that traditional signal processing algorithms often offer very strong baselines. As a result, unlike with image classification, deep methods don't yet offer breakthrough accuracy improvements in this area. However, this is a field of continued and active research. It won't at all be surprising if work on deep signal processing ends up being even more influential than simple image processing in the long run due to the very wide range of applications for such techniques.

It's worth noting that there are many other types of scans doctors use. Given the explosion in deep learning applications powered by strong open source tools, it's a good bet that for each such scan type, there's a study or two attempting to use deep learning for the task. For example, deep learning has been applied to ultrasounds, electrocardiogram (ECG) scans, skin cancer detection, and more.

Convolutional networks are an extraordinarily powerful tool, because so much human activity revolves around processing complex visual information. In addition,

---

the growth of open source frameworks has meant that researchers worldwide have joined the race to apply deep learning techniques on new types of images. In many ways, this type of research is relatively straightforward (on the computational end, at least), as standard tools can be applied without too much fuss. If you're reading this while employed at a company, it's these same properties of deep learning that likely make it interesting to you as a practitioner.

Learning Models as Therapeutics

So far in this chapter, we've seen that learning models can be effective assistants to doctors, helping aid the process of diagnosis and scan understanding. However, there's some exciting evidence that learning models can move past being assistants to doctors to being therapeutic instruments in their own right.

How could this possibly work? One of the greatest powers of deep learning is that it is now feasible for the first time to build practical software that operates on perceptual data. For this reason, machine learning systems could potentially serve as "eyes" and "ears" to differently abled patients. A visual system could help patients with visual impairments more effectively navigate the world. An audio processing system could help patients with hearing impairments more effectively navigate the world. These systems face a number of challenges that other deep models don't, since they have to operate effectively in real time. All the models we've considered so far in this book have been batch systems, suited for deployment on a backend server, not models fit for deployment on a live embedded device. There's a whole host of challenges in dealing with machine learning in production which we won't get into here, but we encourage interested readers to dive into the subject more deeply.

We also note that there's a separate class of software-driven therapeutics that make uses of the powerful effects of modern software on the human brain. A groundswell of recent research has shown that modern software applications such as Facebook, Google, WeChat, and the like can be highly addictive. These apps are designed with bright colors and intended to hit many of the same centers in our brains as casino slot machines. There's growing recognition that digital addictionis a real problem facing many patients.[11] This is a broad area beyond the scope of this book, but we note that there's evidence that this power of modern software can be used for good too. Some software apps have been developed that use the psychological effects of modern apps as therapeutic interventions for patients struggling with depression or other conditions.

---

11 See Digital Addict (*https://en.wikipedia.org/wiki/Digital_addict*) on Wikipedia for more information.

Diabetic Retinopathy

So far in this chapter, we have discussed applications of deep learning to medicine in a theoretical sense. In this section, we'll roll up our sleeves and get our hands dirty with a practical example. In particular, we're going to build a model to help diagnose diabetic retinopathy patient progression.

Diabetic retinopathy is a condition in which diabetes damages the health of the eyes. It is a major cause of blindness, especially in the developing world. The *fundus* is the interior area of the eye that's opposite to the lens. A common strategy for diagnosis of diabetic retinopathy is for doctors to view an image of the patient's fundus and label it manually. Significant work has gone into "fundus photography," which develops techniques to capture patient fundus images (see Figure 8-6).

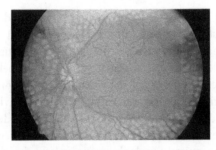

Figure 8-6. An image of a patient fundus from a patient who has undergone scatter laser surgery treatment for diabetic retinopathy. (Source: Wikimedia (https://commons.wiki media.org/wiki/File:Fundus_photo_showing_scatter_laser_surgery_for_diabetic_retin opathy_EDA09.JPG).)

The learning challenge for diabetic retinopathy is to design an algorithm that can classify a patient's disease progress given an image of the patients' fundus. At present, making such predictions requires skilled doctors or technicians. The hope is that a machine learning system could accurately predict disease progression from patient fundus images. This could provide patients with a cheap method of understanding their risk, which they could use before consulting a more expensive expert doctor for a diagnosis.

In addition, unlike EHR data, fundus images don't contain much sensitive information about patients, which makes it easier to gather large fundus image datasets. For these reasons, a number of machine learning studies and challenges have been conducted on diabetic retinopathy datasets. In particular, Kaggle sponsored a contest (*https://www.kaggle.com/c/diabetic-retinopathy-detection*) aimed at creating good diabetic retinopathy models and put together a dataset of high-resolution fundus images. In the remainder of this section, you will learn how to use DeepChem to build a diabetic retinopathy classifier on the Kaggle Diabetic Retinopathy (DR) dataset.

Obtaining the Kaggle Diabetic Retinopathy Dataset

The terms of the Kaggle challenge prohibit us from mirroring the data directly on the DeepChem servers. For this reason, you will need to download the data manually from Kaggle's site. You will have to register an account with Kaggle and download the dataset through their API. The full dataset is quite large (80 GB), so you might choose to download a subset of the data if your internet connection can't handle the full download.

See the GitHub repository (*https://github.com/deepchem/DeepLear ningLifeSciences*) associated with this book for more information on downloading this dataset. The image loading functions here require that the training data is structured in a particular directory structure. Details on this directory format are in the GitHub repo.

The first step to working with this data is to preprocess and load the raw data. In particular, we crop each image to focus on its center square containing the retina. We then resize this center square to be of size 512 by 512.

Dealing with High-Resolution Images

Many image datasets in medicine and science will feature very high-resolution images. While it may be tempting to train deep learning models directly on these high-resolution images, this is usually computationally challenging. One problem is that most modern GPUs have limited memory. That means training very high-resolution models may not be feasible on standard hardware. In addition, most image processing systems (for now) expect their input images to have a fixed shape. This means that high-resolution images from different cameras will have to be cropped to fit within standard shapes.

Luckily, it turns out that cropping and resizing images is usually not terribly damaging to the performance of machine learning systems. It's also common to do more thorough data augmentation, in which a number of perturbed images are automatically generated from each source image. In this particular case study, we performed a few standard data augmentations. We encourage you to dig into the augmentation code since it may prove a useful tool for your own projects.

The core data is stored in a set of directories on disk. We use DeepChem's ImageDataset class to load these images from disk. If you're interested, you can look through this loading and preprocessing code in detail, but we've wrapped it into a convenience helper function. In the style of the MoleculeNet loaders, this function also does a random training, validation, and test split:

```
train, valid, test = load_images_DR(split='random', seed=123)
```

Now that we have the data for this learning task, let's build a convolutional architecture to learn from this dataset. The architecture for this task is fairly standard and resembles other architectures you've already seen in this book, so we don't replicate it here. Here's the invocation of the object wrapper for the underlying convolutional network:

```
# Define and build model
model = DRModel(
    n_init_kernel=32,
    batch_size=32,
    learning_rate=1e-5,
    augment=True,
    model_dir='./test_model')
```

This code sample defines a diabetic retinopathy convolutional network in Deep-Chem. As we will see later, training this model will take some heavy computation. For that reason, we recommend that you download our pretrained model from the Deep-Chem website and use that for your early exploration. We have already trained this model on the full Kaggle Diabetic Retinopathy dataset and stored its weights for your convenience. You can use the following commands to download and store the model (note that the first command should be entered on a single line, with no space after the +/+):

```
wget https://s3-us-west-1.amazonaws.com/deepchem.io/featurized_datasets
    /DR_model.tar.gz
mv DR_model.tar.gz test_model/
cd test_model
tar -zxvf DR_model.tar.gz
cd ..
```

You can then restore the trained model weights as follows:

```
model.restore(checkpoint="./test_model/model-84384")
```

We are restoring a particular pretrained "checkpoint" from this model. We provide more details on the restoration process and the full scripts used to achieve it in the code repository associated with this book. With the pretrained model in place, we can compute some basic statistics upon it:

```
metrics = [
    dc.metrics.Metric(DRAccuracy, mode='classification'),
    dc.metrics.Metric(QuadWeightedKappa, mode='classification')
]
```

There are a number of metrics that are useful for evaluating diabetic retinopathy models. Here we use, DRAccuracy, which is simply the model accuracy (percent of labels which are correct), and Cohen's Kappa, a statistic used to measure agreement between two classifiers. This is useful because the diabetic retinopathy learning task is a multiclass learning problem.

Let's evaluate our pretrained model on the test set with our metrics:

```
model.evaluate(test, metrics, n_classes=5)
```

This produces the following results:

```
computed_metrics: [0.9339595787076572]
computed_metrics: [0.8494075470551462]
```

The basic model gets 93.4% accuracy on our test set. Not bad! (It's important to note that this isn't the same as the Kaggle test set—we've simply partitioned Kaggle's training set into train/valid/test sets for our experimentation. You're welcome to try submitting your trained model to Kaggle for evaluation on their test set, though.) Now, what if you're interested in training the full model from scratch? This will take about a day or two's training on a good GPU system, but is straightforward enough to do:

```
for i in range(10):
  model.fit(train, nb_epoch=10)
  print(model.evaluate(train, metrics, n_classes=5))
  print(model.evaluate(valid, metrics, n_classes=5))
  print(model.evaluate(valid, cm, n_classes=5))
  print(model.evaluate(test, metrics, n_classes=5))
  print(model.evaluate(test, cm, n_classes=5))
```

We train the model for 100 epochs, pausing periodically to print out results from the model. If you're running this job, we recommend making sure that your machine won't shut down or go to sleep halfway through the job. There's nothing as irritating as losing a large job to a sleep screen!

Conclusion

In many ways, the application of machine learning to medicine has the potential to have greater impact than many of the other applications we've seen so far. These other applications may have shifted what you do at work, but machine learning healthcare systems will soon change your personal healthcare experiences, along with the experiences of millions if not billions of others. For this reason, it's worth pausing and thinking through some of the ethical repercussions.

Ethical Considerations

Training data for these systems will likely be biased for the foreseeable future. It's likely that the training data will be drawn from the medical systems of developed economies, and as a result it's possible that the models constructed will be considerably less accurate for portions of the world that currently lack robust medical systems.

In addition, gathering data on patients is itself fraught with potential ethical issues. Medicine has a long and troubled history of experimenting without consent, especially with people from marginalized groups. Consider the case of Henrietta Lacks (*http://rebeccaskloot.com/the-immortal-life/*), an African-American cancer patient in 1950s Baltimore. A cell line cultivated from a tissue sample of Ms. Lacks's tumor ("HeLa") became a standard biological tool and was used in thousands of research papers—yet none of the proceeds from this research ever reached her family. Ms. Lacks's physician did not inform the family of the samples he'd taken, or obtain consent. Her family did not learn about the HeLa cell line till the 1970s, when they were contacted by medical researchers seeking to draw additional samples.

How could this situation repeat itself in the deep learning era? The medical records of a patient could possibly be used to train a learning system without the consent of the patient or their family. Or, perhaps more realistically, the patient or the family could be induced to sign away the rights to their data at the bedside in the hopes of a last-minute cure.

There's something disturbing about these scenarios. None of us would care to learn that our beloved family members' rights have been violated by institutional medicine or profit-seeking startups. How can we seek to prevent these ethical violations from occurring? If you're involved in data gathering efforts, pause and ask where the data is coming from. Were all relevant laws appropriately respected? If you're a scientist or developer at a company or research institution, you will have valuable skills that give you leverage within the organization. If you take a stand, you will influence others in the organization to stand with you. And if your organization refuses to listen, you have valuable skills that will enable you to find a job with an organization that holds itself to high ethical standards.

Job Losses

Most of the fields considered in other chapters in this book are relatively niche scientific disciplines. Thus, the potential of significant advances in the field causing job losses doesn't really exist. Rather, it's to be expected that job growth in these fields will occur as these relatively niche areas will suddenly become accessible to a much wider pool of developers and scientists.

Healthcare and medicine are different. Healthcare is one of the largest industries worldwide, with millions of doctors, nurses, technicians, and more serving the needs of the world's population. What happens as significant fractions of this workforce are confronted with deep learning tools?

Much of medicine is deeply human. Having a trusted primary care provider who you can be sure is looking out for your best interests makes a profound difference to an ill patient. It's very possible that for many patients, care experience could actually improve as much of the busywork is automated out.

In the US, healthcare reform in 2010 (the Affordable Care Act) accelerated the use of EHR systems throughout the American medical system. Many doctors have reported feeling that these EHR systems are deeply unfriendly, requiring many unnecessary administrative actions. Part of this is due simply to poor software design, worsened by regulatory capture that makes it difficult for healthcare institutions to shift to better alternatives. But some of it is due to limitations of present-day software. Use of deep learning systems to allow for more intelligent information handling could lower the burden on doctors, enabling them to spend more time with patients.

In addition, most countries in the world have healthcare systems that don't match those in the United States and Europe. The increasing availability of open source tools and accessible datasets will provide governments and entrepreneurs in the rest of the world the tools they need to serve their constituents.

Summary

In this chapter, you've learned about the history of applying machine learning methods to problems in medicine. We started by giving you an overview of classical methods such as expert systems and Bayesian networks, then shifted into more modern work on electronic health records and medical scans. We ended the chapter with an in-depth case study on training a classifier that predicts diabetic retinopathy patient progression. We also commented in a number of asides about the challenges that learning systems for healthcare face. We'll return to some of these challenges in Chapter 10, where we discuss the interpretability of deep learning systems.

## CHAPTER 9

Generative Models

All the problems we have looked at so far involve, in some way, translating from inputs to outputs. You create a model that takes an input and produces an output. Then you train it on input samples from a dataset, optimizing it to produce the best output for each one.

Generative models are different. Instead of taking a sample as input, they produce a sample as output. You might train the model on a library of photographs of cats, and it would learn to produce new images that look like cats. Or, to give a more relevant example, you might train it on a library of known drug molecules, and it would learn to generate new "drug-like" molecules for use as candidates in a virtual screen. Formally speaking, a generative model is trained on a collection of samples that are drawn from some (possibly unknown, probably very complex) probability distribution. Its job is to produce new samples from that same probability distribution.

In this chapter, we will begin by describing the two most popular types of generative models: *variational autoencoders* and *generative adversarial networks*. We will then discuss a few applications of these models in the life sciences, and work through some code examples.

Variational Autoencoders

An *autoencoder* is a model that tries to make its output equal to its input. You train it on a library of samples and adjust the model parameters so that on every sample the output is as close as possible to the input.

That sounds trivial. Can't it just learn to pass the input directly through to the output unchanged? If that were actually possible it would indeed be trivial, but autoencoders usually have architectures that make it impossible. Most often this is done by forcing the data to go through a bottleneck, as shown in Figure 9-1. For example, the input and output might each include 1,000 numbers, but in between would be a hidden layer containing only 10 numbers. This forces the model to learn how to compress the input samples. It must represent 1,000 numbers worth of information using only 10 numbers.

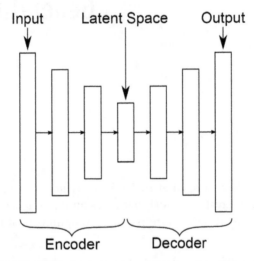

Figure 9-1. Structure of a variational autoencoder.

If the model needed to handle arbitrary inputs, that would be impossible. You can't throw out 99% of the information and still reconstruct the input! But we don't care about arbitrary inputs, only the specific ones in the training set (and others that resemble them). Of all possible images, far less than 1% look anything like cats. An autoencoder doesn't need to work for all possible inputs, only ones that are drawn from a specific probability distribution. It needs to learn the "structure" of that distribution, figure out how to represent the distribution using much less information, and then be able to reconstruct the samples based on the compressed information.

Now let's take the model apart. The middle layer, the one that serves as the bottleneck, is called the *latent space* of the autoencoder. It is the space of compressed representations of samples. The first half of the autoencoder is called the *encoder*. Its job is to take samples and convert them to compressed representations. The second half is called the *decoder*. It takes compressed representations in the latent space and converts them back into the original samples.

This gives us our first clue about how autoencoders could be used for generative modeling. The decoder takes vectors in the latent space and converts them into samples, so we could take random vectors in the latent space (picking a random value for each component of the vector) and pass them through the decoder. If everything goes well, the decoder should produce a completely new sample that still resembles the ones it was trained on.

This sort of works, but not very well. The problem is that the encoder may only produce vectors in a small region of the latent space. If we pick a vector anywhere else in the latent space, we may get an output that looks nothing like the training samples. In other words, the decoder has only learned to work for the particular latent vectors produced by the encoder, not for arbitrary ones.

A variational autoencoder (VAE) adds two features to overcome this problem. First, it adds a term to the loss function that forces the latent vectors to follow a specified distribution. Most often they are constrained to have a Gaussian distribution with a mean of 0 and a variance of 1. We don't leave the encoder free to generate vectors wherever it wants. We force it to generate vectors with a known distribution. That way, if we pick random vectors from that same distribution, we can expect the decoder to work well on them.

Second, during training we add random noise to the latent vector. The encoder converts the input sample to a latent vector, and then we randomly change it a little bit before passing it through the decoder, requiring the output to still be as close as possible to the original sample. This prevents the decoder from being too sensitive to the precise details of the latent vector. If we only change it by a little bit, the output should only change by a little bit.

These changes do a good job of improving the results. VAEs are a popular tool for generative modeling: they produce excellent results on many problems.

Generative Adversarial Networks

A generative adversarial network (GAN) has much in common with a VAE. It uses the same exact decoder network to convert latent vectors into samples (except in a GAN, it is called the *generator* instead of the decoder). But it trains that network in a different way. It works by passing random vectors into the generator and directly evaluating the outputs on how well they follow the expected distribution. Effectively, you create a loss function to measure how well the generated samples match the training samples, then use that loss function to optimize the model.

That sounds simple for a few seconds, until you think about it and realize it isn't simple at all. Could you write a loss function to measure how well an image resembles a cat? No, of course not! You wouldn't know where to begin. So, instead of asking you to come up with that loss function yourself, a GAN learns the loss function from the data.

As shown in Figure 9-2, a GAN consists of two parts. The generator takes random vectors and generates synthetic samples. The second part, called the *discriminator*, tries to distinguish the generated samples from real training samples. It takes a sample as input and outputs a probability that this is a real training sample. It acts as a loss function for the generator.

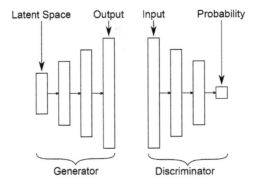

Figure 9-2. Structure of a generative adversarial network.

Both parts are trained simultaneously. Random vectors are fed into the generator, and the output is fed into the discriminator. The parameters of the generator are adjusted to make the discriminator's output as close as possible to 1, while the parameters of the discriminator are adjusted to make its output as close as possible to 0. In addition, real samples from the training set are fed into the discriminator, and its parameters are adjusted to make the output close to 1.

This is the "adversarial" aspect. You can think of it as a competition between the generator and discriminator. The discriminator is constantly trying to get better at distinguishing real samples from fake ones. The generator is constantly trying to get better at fooling the discriminator.

Like VAEs, GANs are a popular type of generative model that produces good results on many problems. The two types of models have distinct strengths and weaknesses. Very roughly speaking, one might say that GANs tend to produce higher-quality samples, while VAEs tend to produce higher-quality distributions. That is, individual samples generated by a GAN will more closely resemble training samples, while the range of samples generated by a VAE will more closely match the range of training samples. Don't take that statement too literally, though. It all depends on the particular problem and the details of the model. Also, countless variations on both approaches have been proposed. There even are models that combine a VAE with a GAN to try to get the best features of both. This is still a very active field of research, and new ideas are published frequently.

Applications of Generative Models in the Life Sciences

Now that we've introduced you to the basics of deep generative models, let's start talking about applications. Broadly speaking, generative models bring a few superpowers to the table. First, they allow for a semblance of "creativity." New samples can be generated according to the learned distribution. This allows for a powerful complement to a creative process that can tie into existing efforts in drug or protein design. Second, being able to model complex systems accurately with generative models could allow scientists to build an understanding of complex biological processes. We'll discuss these ideas in more depth in this section.

Generating New Ideas for Lead Compounds

A major part of a modern drug discovery effort is coming up with new compounds. This is mostly done semiannually, with expert human chemists suggesting modifications to core structures. Often, this will involve projecting a picture of the current molecular series on a screen and having a room full of senior chemists suggest modifications to the core structure of the molecule. Some subset of these suggested molecules are actually synthesized and tested, and the process repeats until a suitable molecule is found or the program is dropped. This process has powerful advantages since it can draw upon the deep intuition of expert chemists who may be able to identify flaws with a potential structure (perhaps it resembles a compound they've seen before which caused unexplained liver failure in rats) that may not be easy to identify algorithmically.

At the same time, though, this process is very human-limited. There aren't that many talented and experienced senior chemists in the world, so the process can't scale outward. In addition, it makes it very challenging for a pharmaceutical division in a country that has historically lacked drug discovery expertise to bootstrap itself. A generative model of molecular structures could serve to overcome these limitations. If the model were trained on a suitable molecular representation, it might be able to rapidly suggest new alternative compounds. Access to such a model could help improve current processes by suggesting new chemical directions that may have been missed by human designers. It's worth noting that such design algorithms have serious caveats, though, as we will see a little later in this chapter.

Protein Design

Design of new enzymes and proteins is a major business these days. Engineered enzymes are used widely in modern manufacturing. (There's a good chance your laundry detergent holds some enzymes!) However, in general, design of new enzymes has proven challenging. Some early work has shown that deep models can have some success at predicting protein function from sequence. It's not unreasonable at all to envision using deep generative models to suggest new protein sequences that might have desired properties.

The introduction of generative models for this purpose could be even more impactful than for small molecule design. Unlike with small molecules, it can be very tricky for human experts to predict the downstream effects of mutations to a given protein.

Using generative models can allow for richer protein design, enabling directions beyond the capability of human experts today.

A Tool for Scientific Discovery

Generative models can be a powerful tool for scientific discovery. For example, having an accurate generative model of a tissue development process (*https://www.ncbi.nlm.nih.gov/pmc/articles/PMC6119234/*) could be extremely valuable to developmental biologists or as a tool in basic science. It might be possible to create "synthetic assays" where we can study tissue development in many combinations of environmental conditions by using the generative model to run rapid simulations. This future is still a ways off, since we'd need generative models that work effectively as initial conditions change. This will take some more research beyond the current state of the art. Nevertheless, the vision is exciting because generative modeling could allow for biologists to build effective models of extremely complex developmental and physiological processes and test their hypotheses of how these systems evolve.

The Future of Generative Modeling

Generative models are challenging! The first GANs were only capable of generating blurry images that were barely recognizable as faces. The latest GANs (at the time of writing) are capable of generating images of faces that are more or less indistinguishable from true photographs. It is likely that in the next decade, these models will be further refined to allow for generative videos. These developments will have profound repercussions on modern societies. For much of the last century, photographs have been routinely used as "proof" of crimes, quality, and more. As generative tools develop, this standard of proof will fall short, as arbitrary images will be able to be "photoshopped." This development will pose a major challenge for criminal justice and even international relations.

At the same time, it's likely that the advent of high-fidelity generative video will trigger a revolution in modern science. Imagine high-quality generative models of embryonic development! It might be feasible to model the effects of CRISPR genetic modifications or understand developmental processes in greater detail than has ever been possible. Improvements in generative models will have effects in other fields of science too. It's likely that generative modeling will become a powerful tool in physics and climate science, allowing for more powerful simulations of complex systems. However, it's worth emphasizing that these improvements today remain in the future; much basic science has to be done to mature these models to useful stability.

Working with Generative Models

Now let's work through a code example. We will train a VAE to generate new molecules. More specifically, it will output SMILES strings. This choice of representation has distinct advantages and disadvantages compared to some of the other representations we have discussed. On the one hand, SMILES strings are very simple to work with. Each one is just a sequence of characters drawn from a fixed alphabet. That allows us to use a very simple model to process them. On the other hand, SMILES strings are required to obey a complex grammar. If the model does not learn all the subtleties of the grammar, then most of the strings it produces will be invalid and not correspond to any molecule.

The first thing we need is a collection of SMILES strings on which to train the model. Fortunately, MoleculeNet provides us with lots to choose from. For this example, we will use the MUV dataset. The training set includes 74,469 molecules of varying sizes and structures. Let's begin by loading it:

```
import deepchem as dc
tasks, datasets, transformers = dc.molnet.load_muv()
train_dataset, valid_dataset, test_dataset = datasets
train_smiles = train_dataset.ids
```

Next, we need to define the vocabulary our model will work with. What is the list of characters (or "tokens") that can appear in a string? How long are strings allowed to be? We can determine these from the training data by creating a sorted list of every character that appears in any training molecule:

```
tokens = set()
for s in train_smiles:
  tokens = tokens.union(set(s))
tokens = sorted(list(tokens))
max_length = max(len(s) for s in train_smiles)
```

Now we need to create a model. What sort of architecture should we use for the encoder and decoder? This is an ongoing field of research. Various papers have been published suggesting different models. For this example, we will use DeepChem's AspuruGuzikAutoEncoder class, which implements a particular published model. It uses a convolutional network for the encoder and a recurrent network for the decoder. You can consult the original paper (*https://arxiv.org/abs/1610.02415*) if you are interested in the details, but they are not necessary to follow the example. Also notice that we use ExponentialDecay for the learning rate. The rate is initially set to 0.001, then decreased by a little bit (multiplied by 0.95) after every epoch. This helps optimization to proceed more smoothly in many problems:

```
from deepchem.models.optimizers import ExponentialDecay
from deepchem.models.seqtoseq import AspuruGuzikAutoEncoder
batch_size = 100
```

```
batches_per_epoch = len(train_smiles)/batch_size
learning_rate = ExponentialDecay(0.001, 0.95, batches_per_epoch)
model = AspuruGuzikAutoEncoder(tokens, max_length, model_dir='vae',
                               batch_size=batch_size,
                               learning_rate=learning_rate)
```

We are now ready to train the model. Instead of using the standard `fit()` method that takes a `Dataset`, `AspuruGuzikAutoEncoder` provides its own `fit_sequences()` method. It takes a Python generator object that produces sequences of tokens (SMILES strings in our case). Let's train for 50 epochs:

```
def generate_sequences(epochs):
  for i in range(epochs):
    for s in train_smiles:
      yield (s, s)

model.fit_sequences(generate_sequences(50))
```

If everything has gone well, the model should now be able to generate entirely new molecules. We just need to pick random latent vectors and pass them through the decoder. Let's create a batch of one thousand vectors, each of length 196 (the size of the model's latent space).

As noted previously, not all outputs will actually be valid SMILES strings. In fact, only a small fraction of them are. Fortunately, we can easily use RDKit to check them and filter out the invalid ones:

```
import numpy as np
from rdkit import Chem
predictions = model.predict_from_embeddings(np.random.normal(size=(1000,196)))
molecules = []
for p in predictions:
  smiles = ''.join(p)
  if Chem.MolFromSmiles(smiles) is not None:
    molecules.append(smiles)
for m in molecules:
  print(m)
```

Analyzing the Generative Model's Output

In addition to the problem of invalid outputs, many of the molecules corresponding to the SMILES strings that are output may not be characteristic of drug molecules. So, we need to develop strategies that will enable us to quickly identify molecules that are not drug-like. These strategies can best be explained through a practical example. Let's assume that this is the list of SMILES strings that came from our generative model:

```
smiles_list = ['CCCCCCNNNCCOCC',
'O=C(O)C(=O)ON/C=N/CO',
'C/C=N/COCCNSCNCCNN',
```

```
'CCCNC(C(=O)O)c1cc(OC(OC)[SH](=O)=O)ccc1N',
'CC1=C2C=CCC(=CC(Br)=CC=C1)C2',
'CCN=NNNC(C)OOCOOOOOCOOO',
'N#CNCCCCCOCCOC1COCNN1CCCCCCCCCCCCCCCCCCCOOOOOSNNCCCCCSCSCCCCCCCCCOCOOOSS',
'CCCC(=O)NC1=C(N)C=COO1',
'CCCSc1cc2nc(C)cnn2c1NC',
'CONCN1N=NN=CC=C1CC1SSS1',
'CCCOc1ccccc1OSNNOCCNCSNCCN',
'C[SH]1CCCN2CCN2C=C1N',
'CC1=C(C#N)N1NCCC1=COOO1',
'CN(NCNNNN)C(=O)CCSCc1ccco1',
'CCCN1CCC1CC=CC1=CC=S1CC=O',
'C/N=C/c1ccccc1',
'Nc1cccooo1',
'CCOc1ccccc1CCCNC(C)c1nccs1',
'CNNNNNNc1nocc1CCNNC(C)C',
'COC1=C(CON)C=C2C1=C(C)c1ccccc12',
'CCOCCCCNN(C)C',
'CCCN1C(=O)CNC1C',
'CCN',
'NCCNCc1cccc2c1C=CC=CC=C2',
'CCCCCN(NNNCNCCCCCCCCCCSCCCCCCCCCCCCCCCCNCCNCCCCSSCSSSSSSCCCCCCCCCCCCCCCSCCCCCSC)\
C(O)OCCN',
'CCCS1=CC=C(C)N(CN)C2NCC2=C1',
'CCNCCCCCCOc1cccc(F)c1',
'NN1O[SH](CCCCO)C12C=C2',
'Cc1cc2cccc3c(CO)cc-3ccc-2c1']
```

The first step in our analysis will be to examine the molecules and determine whether there are any that we want to discard. We can use some of the facilities in RDKit, which is included as part of DeepChem, to examine the molecules represented by these strings. In order to evaluate the strings, we must first convert them to molecule objects. We can do this using the following list comprehension:

```
molecules = [Chem.MolFromSmiles(x) for x in smiles_list]
```

One factor we may want to examine is the size of the molecules. Molecules with fewer than 10 atoms are unlikely to generate sufficient interaction energy to produce a measurable signal in a biological assay. Conversely, molecules with more than 50 atoms may not be capable of dissolving in water and may create other problems in biological assays. We can get a rough estimate of the sizes of the molecules by calculating the number of non-hydrogen atoms in each molecule. The following code creates a list of the number of atoms in each molecule. For convenience, we sort the array so that we can more easily understand the distribution (if we had a larger list of molecules we would probably want to generate a histogram for this distribution):

```
print(sorted([x.GetNumAtoms() for x in molecules]))
```

The results are as follows:

```
[3, 8, 9, 10, 11, 11, 12, 12, 13, 14, 14, 14, 15,
 16, 16, 16, 17, 17, 17, 17, 18, 19, 19, 20, 20, 22, 24, 69, 80]
```

We can see that there are four very small molecules as well as two large molecules. We can use another list comprehension to remove molecules with 10 or fewer than 50 atoms:

```
good_mol_list = [x for x in molecules if x.GetNumAtoms() > 10
        and x.GetNumAtoms() < 50]
print(len(good_mol_list))
23
```

This list comprehension reduces our previous list of 29 molecules to 23.

In practice, we can use a number of other calculated properties to evaluate the quality of the generated molecules. Several recent generative model publications use calculated molecular properties to determine which of the generated molecules to retain or discard. One of the more common methods for determining whether molecules are similar to known drugs, or "drug-like," is known as the quantitative estimate of drug-likeness (QED). The QED metric, which was originally published by Bickerton and coworkers,[1] scores molecules by comparing a set of properties calculated for each molecule with distributions of the same properties in marketed drugs. This score ranges between 0 and 1, with values closer to 1 being considered more drug-like.

We can use RDKit to calculate QED values for our remaining molecules and retain only those molecules with QED > 0.5 as follows:

```
qed_list = [QED.qed(x) for x in good_mol_list]
final_mol_list = [(a,b) for a,b in
        zip(good_mol_list,qed_list) if b > 0.5]
```

As our final step, we can visualize the chemical structures of final_mol_list and the corresponding QED scores:

```
MolsToGridImage([x[0] for x in final_mol_list],
molsPerRow=3,useSVG=True,
subImgSize=(250, 250),
legends=[f"{x[1]:.2f}" for x in final_mol_list])
```

The results are shown in Figure 9-3.

---

1 Bickerton, Richard G. et al. "Quantifying the Chemical Beauty of Drugs." *http://dx.doi.org/10.1038/nchem.1243.* 2012.

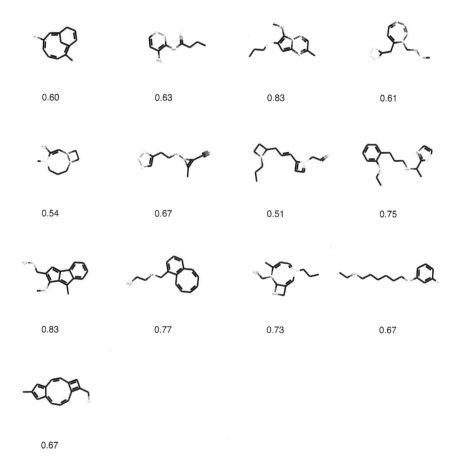

0.60   0.63   0.83   0.61

0.54   0.67   0.51   0.75

0.83   0.77   0.73   0.67

0.67

Figure 9-3. Chemical structures of the generated molecules along with their QED scores.

While these structures are valid and have reasonably high QED scores, they still contain functionality that may be chemically unstable. Strategies for identifying and removing problematic molecules like these are discussed in the next section.

Conclusion

While generative models provide an interesting means of producing ideas for new molecules, some key issues still need to be resolved to ensure their general applicability. The first is ensuring that the generated molecules will be chemically stable and that they can be physically synthesized. One current method to assess the quality of molecules produced by a generative model is to observe the fraction of the generated molecules that obey standard rules of chemical valence—in other words, ensuring that each carbon atom has four bonds, each oxygen atom has two bonds, each fluorine atom has one bond, and so on. These factors become especially important when decoding from a latent space with a SMILES representation. While a generative model may have learned the grammar of SMILES, there may be nuances that are still missing.

The fact that a molecule obeys standard rules of valence does not necessarily ensure that it will be chemically stable. In some cases, a generative model may produce molecules containing functional groups that are known to readily decompose. As an example, consider the molecule in Figure 9-4. The functional group highlighted in the circle, known as a hemiacetal, is known to readily decompose.

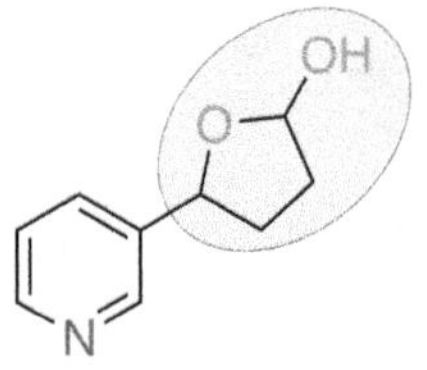

Figure 9-4. A molecule containing an unstable group.

In practice, the probability of this molecule existing and being chemically stable is very small. There are dozens of chemical functionalities like this which are known to be unstable or reactive. When synthesizing molecules in a drug discovery project, medicinal chemists know to avoid introducing these functional groups. One way of imparting this sort of "knowledge" to a generative model is to provide a set of filters that can be used to postprocess the model output and remove molecules that may be problematic. In Chapter 11, we will provide a further discussion of some of these filters and how they are used in virtual screening. Many of the same techniques used to identify potentially problematic screening compounds can also be used to evaluate virtual molecules that are created by a generative model.

In order to test the biological activity of a molecule produced by a generative model, that molecule must first be synthesized by a chemist. The science of organic chemical synthesis has a rich history going back more than one hundred years. In this time, chemists have developed thousands of chemical reactions to synthesize drugs and drug-like molecules. The synthesis of a drug-like molecule typically requires somewhere between 5 and 10 chemical reactions, often referred to as "steps." While some drug-like molecules can be readily synthesized, the synthetic route to more complex drug molecules may require more than 20 steps. Despite more than 50 years of work on automating the planning of organic syntheses, much of the process is still driven by human intuition followed by trial and error.

Fortunately, recent developments in deep learning are providing new ways of planning the synthesis of drug-like molecules. A number of groups have published methods that use deep learning to propose routes that can be used to synthesize molecules.

As input, the model is given a molecule, often referred to as a *product*, and the set of steps that were used to synthesize that molecule. By training with thousands of product molecules and the steps used for synthesis, a deep neural network is able to learn the relationship between product molecules and reaction steps. When presented with a new molecule, the model suggests a set of reactions that could be used to synthesize the molecule. In one test, the synthetic routes produced by these models were presented to human chemists for evaluation. These evaluators felt that the routes generated by the models were comparable in quality to routes generated by human chemists.

The application of deep learning to organic synthesis is a relatively new field. It is hoped that the field will continue to evolve and that these models become an important tool for organic chemists. One can imagine a day in the not too distant future where these synthesis planning capabilities could be paired with robotic automation to create a fully automated platform. However, there are difficulties to overcome.

One potential roadblock in the broad adoption of deep learning in organic synthesis is data availability. The majority of the information used to train these models is in databases which are the property of a small number of organizations. If these organizations decide to only utilize this data for their internal efforts, the field will be left with very few alternatives.

Another factor that may limit the advance of generative models is the quality of the predictive models that are used to drive molecule generation. Regardless of the architecture used to develop a generative model, some function must be used to evaluate the generated molecules and to direct the search for new molecules. In some cases, we may be able to develop reliable predictive models. In other cases, the models may be less reliable. While we can test our models on external validation sets, it is often difficult to determine the scope of a predictive model. This scope, also known as the "domain of applicability," is the degree to which one can extrapolate outside the molecules on which a model was trained. This applicability domain is not well defined, so it may be difficult to determine how well a model will work on novel molecules produced by a generative model.

Generative models are a relatively new technique, and it will be interesting to see how this field evolves in the coming years. As our ability to use deep learning to predict routes for organic synthesis and build predictive models improves, the power of generative models will continue to grow.

## CHAPTER 10

Interpretation of Deep Models

At this point we have seen lots of examples of training deep models to solve problems. In each case we collect some data, build a model, and train it until it produces the correct outputs on our training and test data. Then we pat ourselves on the back, declare the problem to be solved, and go on to the next problem. After all, we have a model that produces correct predictions for input data. What more could we possibly want?

But often that is only the beginning! Once you finish training the model there are lots of important questions you might ask. How does the model work? What aspects of an input sample led to a particular prediction? Can you trust the model's predictions? How accurate are they? Are there situations where it is likely to fail? What exactly has it "learned"? And can it lead to new insights about the data it was trained on?

All of these questions fall under the topic of *interpretability*. It covers everything you might want from a model beyond mechanically using it to make predictions. It is a very broad subject, and the techniques it encompasses are as diverse as the questions they try to answer. We cannot hope to cover all of them in just one chapter, but we will try to at least get a taste of some of the more important approaches.

To do this, we will revisit examples from earlier chapters. When we saw them before, we just trained models to make predictions, verified their accuracy, and then considered our work complete. Now we will take a deeper look and see what else we can learn.

Explaining Predictions

Suppose you have trained a model to recognize photographs of different kinds of vehicles. You run it on your test set and find it accurately distinguishes between cars, boats, trains, and airplanes. Does that make it ready to put into production? Can you trust it to keep producing accurate results in the future?

Maybe, but if wrong results lead to serious consequences you might find yourself wishing for some further validation. It would help if you knew *why* the model produced its particular predictions. Does it really look at the vehicle, or is it actually relying on unrelated aspects of the image? Photos of cars usually also include roads. Airplanes tend to be silhouetted against the sky. Pictures of trains usually include tracks, and ones of boats include lots of water. If the model is really identifying the background rather than the vehicle, it may do well on the test set but fail badly in unexpected cases. A boat silhouetted against the sky might be classified as an airplane, and a car driving past water might be identified as a boat.

Another possible problem is that the model is fixating on overly specific details. Perhaps it does not really identify pictures of *cars*, just pictures that include *license plates*. Or perhaps it is very good at identifying life preservers, and has learned to associate them with pictures of boats. This will usually work, but will fail when shown a car driving past a swimming pool with a life preserver visible in the background.

Being able to explain why the model made a prediction is an important part of interpretability. When the model identifies a photograph of a car, you want to know that it made the identification based on the actual car, not based on the road, and not based on only one small part of the car. In short, you want to know that it gave the right answer *for the right reasons*. That gives you confidence that it will also work on future inputs.

As a concrete example, let's return to the diabetic retinopathy model from Chapter 8. Recall that this model takes an image of a retina as input, and predicts the presence and severity of diabetic retinopathy in the patient. Between the input and output are dozens of Layer objects and more than eight million trained parameters. We want to understand why a particular input led to a particular output, but we cannot hope to learn that just by looking at the model. Its complexity is far beyond human comprehension.

Many techniques have been developed for trying to answer this question. We will apply one of the simplest ones, called *saliency mapping*.[1] The essence of this technique is to ask which pixels of the input image are most important (or "salient") for determining the output. In some sense, of course, *every* pixel is important. The output is a hugely complex nonlinear function of all the inputs. In the right image, any pixel might contain signs of disease. But in a particular image only a fraction of them do, and we want to know which ones they are.

---

1 Simonyan, K., A. Vedaldi, and A. Zisserman. "Deep Inside Convolutional Networks: Visualising Image Classification Models and Saliency Maps." Arxiv.org (*https://arxiv.org/abs/1312.6034*). 2014.

Saliency mapping uses a simple approximation to answer this question: just take the derivative of the outputs with respect to all the inputs. If a region of the image contains no sign of disease, small changes to any individual pixel in that region should have little effect on the output. The derivative should therefore be small. A positive diagnosis involves correlations between many pixels. When those correlations are absent, they cannot be created just by changing one pixel. But when they are present, a change to any one of the participating pixels can potentially strengthen or weaken the result. The derivative should be largest in the "important" regions the model is paying attention to.

Let's look at the code. First we need to build the model and reload the trained parameter values:

```
import deepchem as dc
import numpy as np
from model import DRModel
from data import load_images_DR

train, valid, test = load_images_DR(split='random', seed=123)
model = DRModel(n_init_kernel=32, augment=False, model_dir='test_model')
model.restore()
```

Now we can use the model to make predictions about samples. For example, let's check the predictions for the first 10 test samples:

```
X = test.X
y = test.y
for i in range(10):
  prediction = np.argmax(model.predict_on_batch([X[i]]))
  print('True class: %d, Predicted class: %d' % (y[i], prediction))
```

Here is the output:

```
True class: 0, Predicted class: 0
True class: 2, Predicted class: 2
True class: 0, Predicted class: 0
True class: 0, Predicted class: 0
True class: 3, Predicted class: 0
True class: 2, Predicted class: 2
True class: 0, Predicted class: 0
True class: 0, Predicted class: 0
True class: 0, Predicted class: 0
True class: 2, Predicted class: 2
```

It gets 9 of the first 10 samples right, which is not bad. But what is it looking at when it makes its predictions? Saliency mapping can give us an answer. DeepChem makes this easy:

```
saliency = model.compute_saliency(X[0])
```

`compute_saliency()` takes the input array for a particular sample and returns the derivative of every output with respect to every input. We can get a better sense of what this means by looking at the shape of the result:

```
print(saliency.shape)
```

This reports it is an array of shape `(5, 512, 512, 3)`. `X[0]` is the 0th input image, which is an array of shape `(512, 512, 3)`, the last dimension being the three color components. In addition, the model has five outputs, the probabilities of the sample belonging to each of the five classes. `saliency` contains the derivative of each of the five outputs with respect to each of the 512×512×3 inputs.

This needs a little processing to be made more useful. First, we want to take the absolute value of every element. We don't care whether a pixel should be made darker or lighter to increase the output, just that it has an effect. Then we want to condense it down to just one number per pixel. That could be done in various ways, but for now we will simply sum over the first and last dimensions. If any color component affects any of the output predictions, that makes the pixel important. Finally, we will normalize the values to be between 0 and 1:

```
sal_map = np.sum(np.abs(saliency), axis=(0,3))
sal_map -= np.min(sal_map)
sal_map /= np.max(sal_map)
```

Let's see what it looks like. Figure 10-1 shows a sample that the model correctly identifies as having severe diabetic retinopathy. The input image is on the left, and the right side highlights the most salient regions in white.

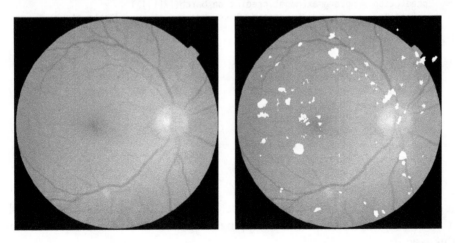

Figure 10-1. Saliency map for an image with severe diabetic retinopathy.

The first thing we notice is that the saliency is widely spread over the whole retina, not just in a few spots. It is not uniform, however. Saliency is concentrated along the blood vessels, and especially at points where blood vessels branch. Indeed, some of the indications a doctor looks for to diagnose diabetic retinopathy include abnormal blood vessels, bleeding, and the growth of new blood vessels. The model appears to be focusing its attention on the correct parts of the image, the same ones a doctor would look at most closely.

Optimizing Inputs

Saliency mapping and similar techniques tell you what information the model was focusing on when it made a prediction. But how exactly did it interpret that information? The diabetic retinopathy model focuses on blood vessels, but what does it look for to distinguish healthy from diseased blood vessels? Similarly, when a model identifies a photograph of a boat, it's good to know it made the identification based on the pixels that make up the boat, not the ones that make up the background. But what about those pixels led it to conclude it was seeing a boat? Was it based on color? On shape? Combinations of small details? Could there be unrelated pictures the model would equally confidently (but incorrectly) identify as a boat? What exactly does the model "think" a boat looks like?

A common approach to answering these questions is to search for inputs that maximize the prediction probability. Out of all possible inputs you could put into the model, which ones lead to the strongest predictions? By examining those inputs, you can see what the model is really "looking for." Sometimes it turns out to be very different from what you expect! Figure 10-2 shows images that have been optimized to produce strong predictions when fed into a high-quality image recognition model. The model identifies each image as the listed category with very high confidence, yet to a human they have almost no resemblance!

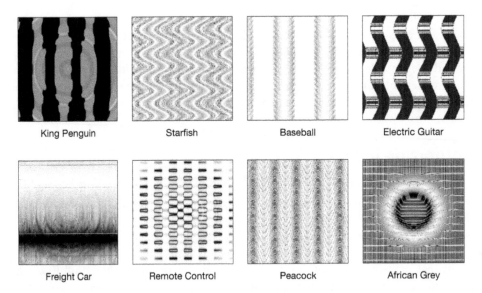

| King Penguin | Starfish | Baseball | Electric Guitar |

| Freight Car | Remote Control | Peacock | African Grey |

Figure 10-2. Images that fool a high-quality image recognition model. (Source: Arxiv.org (https://arxiv.org/abs/1412.1897).)

As an example, consider the transcription factor binding model from Chapter 6. Recall that this model takes a DNA sequence as input, and predicts whether the sequence contains a binding site for the transcription factor JUND. What does it think a binding site looks like? We want to consider all possible DNA sequences and find the ones for which the model most confidently predicts that a binding site is present.

Unfortunately, we can't really consider all possible inputs. There are 4^{101} possible DNA sequences of length 101. If you needed only one nanosecond to examine each one, it would take many times longer than the age of the universe to get through all of them. Instead, we need a strategy to sample a smaller number of inputs.

One possibility is just to look at the sequences in the training set. In this case, that is actually a reasonable strategy. The training set covers tens of millions of bases from a real chromosome, so it likely is a good representation of the inputs that will be used with this model in practice. Figure 10-3 shows the 10 sequences from the training set for which the model produces the highest output. Each of them is predicted to have a binding site with greater than 97% probability. Nine of them do in fact have binding sites, while one is a false positive. For each one, we have used saliency mapping to identify what the model is focusing on and colored the bases by their saliency.

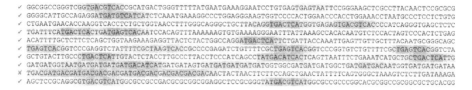

✓ GGCGGCCGGGTCGGTGACGTCACCGCATGACTGGGTTTTTATGAATGAAAGGAATCCTGTGAGTGAGTAATTCCGGGAAGCTCGCCTTACAACTCCGCGCG
✓ GGGGCATTGCCAGAGGATGATGTCATCATCTCAAATGAAAGGCCCTGGAGGGAAGTGGTCCCCACTGGAACCCACCTGGAAACCTAATGCCCTCCTCTGTG
✓ CTGAATGAACACCAAGGTCACCTCTGCTGGTAACCTTTGGGCAGGGCTGCTTACAGGTGACTCATGGTGAGAGTGACGTCACCCATCAGGGTGAGCTCTC
✓ TGATTTCATGACTCACTGATGAGTCACAATCCACAGTTTAAAAAAGTGTGAAAAGGGAATTTATTAAAGCCACACAATGTCTCCACTAGTCCCACTCTGAG
✓ ACATTTTGCTCTTCTCAGCTGGTAAGAAAGAGGTTACTCTACTGGCCAGGATGACTCATTCTGATTACCAAATTGAGTTGTTGCTTTACAATGCGGGCAGC
✓ TGAGTCACGGTCCCGAGGTCTATTTTCGCTAAGTCACCGCCCCGAGATCTGTTTTCGCTGAGTCACGGTCCCGGTGTCTGTTTTCGCTGAGTCACGGTCTA
✓ GCTGTACTTGCCCTGACTCATTGTACTCTACCTTGCCCTTACCTCCCATCAGCCTATGACATCACTCAGTTAATTTCTGAAATCATGCTGCTGACTCATTG
✓ GATGATGGTAATGATGATGATGATGACATCATGATGATAGTGATGATGATGATGGTGGCGATGATGATGGCTGATGATGACAATGGTGATGATGATAA
✗ TGACGATGACGATGACGACGACGATGACGACGACGACGACGACAACTACTAACTTCTTCCAGCTGAACTATTTTCAGTGGGCTAAAGTCTCTTGATAAAGA
✓ AGCTCCGCAGGCGTGACGTCATGGCGCCGCCGACGCGCGGCGGAGGCTCCGCGGGTATGACGTCATGGCGCCGCCCGGCACGCGGCCGCGGCGCTGCACGG

Figure 10-3. The 10 training examples with the highest predicted outputs. Checkmarks indicate the samples that contain actual binding sites.

Looking at these inputs, we can immediately see the core pattern it is recognizing: TGA ... TCA, where ... consists of one or two bases that are usually C or G. The saliency indicates it also pays some attention to another one or two bases on either side. The previous base can be an A, C, or G, and the following base is always either a C or T. This agrees with the known binding motif for JUND, which is shown in Figure 10-4 as a position weight matrix.

Figure 10-4. The known binding motif for JUND, represented as a position weight matrix. The height of each letter indicates the probability of that base appearing at the corresponding position.

The one sequence that was incorrectly predicted to have a binding site does not contain this pattern. Instead, it has several repetitions of the pattern TGAC, all close together. This looks like the beginning of a true binding motif, but it is never followed by TCA. Apparently our model has learned to identify the true binding motif, but it also can be misled when several incomplete versions occur in close proximity.

The training samples will not always be a good representation of the full range of possible inputs. If your training set consists entirely of photographs of vehicles, it tells you nothing about how the model responds to other inputs. Perhaps if shown a photograph of a snowflake, it would confidently label it as a boat. Perhaps there even are inputs that look nothing like photographs—maybe just simple geometric patterns or even random noise—that the model would identify as boats. To test for this possibility, we can't rely on the inputs we already have. Instead, we need to let the model tell us what it is looking for. We start with a completely random input, then use an optimization algorithm to modify it in ways that increase the model's output.

Let's try doing this for the TF binding model. We begin by generating a completely random sequence and computing the model's prediction for it:

```
best_sequence = np.random.randint(4, size=101)
best_score =
    float(model.predict_on_batch([dc.metrics.to_one_hot(best_sequence, 4)]))
```

Now to optimize it. We randomly select a position within the sequence and a new base to set it to. If this change causes the output to increase, we keep it. Otherwise, we discard the change and try something else:

```
for step in range(1000):
    index = np.random.randint(101)
    base = np.random.randint(4)
    if best_sequence[index] != base:
      sequence = best_sequence.copy()
      sequence[index] = base
      score = float(model.predict_on_batch([dc.metrics.to_one_hot(sequence, 4)]))
      if score > best_score:
        best_sequence = sequence
        best_score = score
```

This rapidly leads to sequences that maximize the predicted probability. Within 1,000 steps, we usually find the output has saturated and equals 1.0.

Figure 10-5 shows 10 sequences generated by this process. All instances of the three most common binding patterns (TGACTCA, TGAGTCA, and TGACGTCA) are highlighted. Every sequence contains at least one occurrence of one of these patterns, and usually three or four. Sequences that maximize the model's output have exactly the properties we expect them to, which gives us confidence that the model is working well.

```
GACGTCATCCCTTACGATGACGTCATCATAACGGCGACGATGACTCTACTGATGAGTCATCGCTGTGACGACGTTACTGCCGCTGACGCAATTGATGACGT
CGCGGCGACGGTTCCGATTACTCATCGGGTGATGACGTCGTGACGTCATCGGTGATGACGACGTCACCTCCGGAACGGTGACGACGGTGATGACGTAACTC
CGCTGCGGTGATGACTCATTCCGTTGTGAGTCATCGCTAACGGTTCCGAAAGTTTTCCGACGGCGATGAGTCATCGGTGACGATGACGATGACTCATTGTA
CGTCATTCGTGAGTCATGACTTATCGCGAAATGATGACTCCGTTCCGATGACTCATCGGCGCCGTCCGGTCAGGAATGACGTCATGATGACTCATTACGTC
GATGACTCATCTATGACTCAATGACGTCATTCGCGTGATTACGTCCTTTATAACGATGACGTCATCACTCACGGAACCGATCCGATTACGACCGGCGCTCC
TCAGTGATGACTGAGTCATCATGACGTCATCGCGTTCCGGTTACGTCGAGTACCGCGGCGGATGACGTCATCGCAGACCGGGGATGACTTCACTGGTGTGA
GTCACCTCGAAACGGTACGGCTCTGAGAGTACGTCACCGAGTATCCGATCCGGGCGATGACTCATCGTTCGGTGATGATGATGTCATCATCGATGACTCAT
GTCACGGGTAATGATGACGTCATCGCTTCGGATTGCCCGTCCGGAACGATGACGTCATATCGGTTATGACGTCAAGACGTCATCGCTTGCAGATGATGACG
CCTGCGATGACGTCATCGATTAGCATCGATGACGTCACTTACTGCTCCGACGATGACTCAATGAGTCATCATCAATGGTGACGACCGGGCCGGTTACGGTT
AGCGCTCCGTCCGGAATGATGACGTCACTTTGGTGACTCAGTAACTGTCCGCACCGATGACTCATCGGGTACGGTGAGTCATTGCACTGGTACGACATCTA
```

Figure 10-5. Example sequences that have been optimized to maximize the model's output.

Predicting Uncertainty

Even when you have convinced yourself that a model produces accurate predictions, that still leaves an important question: exactly *how* accurate are they? In science, we are rarely satisfied with just a number; we want an uncertainty for every number. If the model outputs 1.352, should we interpret that as meaning the true value is between 1.351 and 1.353? Or between 0 and 3?

As a concrete example, we will use the solubility model from Chapter 4. Recall that this model takes a molecule as input, represented as a molecular graph, and outputs a number indicating how easily it dissolves in water. We built and trained the model with the following code.

```
tasks, datasets, transformers = dc.molnet.load_delaney(featurizer='GraphConv')
train_dataset, valid_dataset, test_dataset = datasets
model = GraphConvModel(n_tasks=1, mode='regression', dropout=0.2)
model.fit(train_dataset, nb_epoch=100)
```

When we first examined this model, we evaluated its accuracy on the test set and declared ourselves satisfied. Now let's try to do a better job of quantifying its accuracy.

A very simple thing we might try doing is just to compute the root-mean-squared (RMS) error of the model's predictions on the test set:

```
y_pred = model.predict(test_dataset)
print(np.sqrt(np.mean((test_dataset.y-y_pred)**2)))
```

This reports an RMS error of 0.396. Should we therefore use that as the expected uncertainty in all predictions made by the model? *If* the test set is representative of all inputs the model will be used on, and *if* all errors follow a single distribution, that might be a reasonable thing to do. Unfortunately, neither of those is a safe assumption! Some predictions may have much larger errors than others, and depending on the particular molecules that happen to be in the test set, their average error might be either higher or lower than what you will encounter in practice.

We really want to associate a different uncertainty with every output. We want to know in advance which predictions are more accurate and which are less accurate. To do that, we need to consider more carefully the multiple factors that contribute to errors in a model's predictions.[2] As we will see, there are two fundamentally different types of uncertainty that must be included.

Figure 10-6 shows the true versus predicted solubilities of the molecules in the training set. The model is doing a very good job of reproducing the training set, but not a perfect job. The points are distributed in a band with finite width around the diagonal. Even though it was trained on those samples, the model still has some error when predicting them. Given that, we have to expect it to have *at least* as much error on other data it was not trained on.

---

2 Kendall, A., and Y. Gal, "What Uncertainties Do We Need in Bayesian Deep Learning for Computer Vision?" *https://arxiv.org/abs/1703.04977*. 2017.

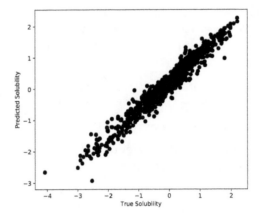

Figure 10-6. True versus predicted solubilities for the molecules in the training set.

Notice that we are only looking at the training set. This uncertainty can be determined entirely from information that is available at training time. That means we can train a model to predict it! We can add another set of outputs to the model: for every value it predicts, it will also output an estimate of the uncertainty in that prediction.

Now consider Figure 10-7. We have repeated the training process 10 times, giving us 10 different models. We have used each of them to predict the solubility of 10 molecules from the test set. All of the models were trained on the same data, and they have similar errors on the training set, yet they produce different predictions for the test set molecules! For each molecule, we get a range of different solubilities depending on which model we use.

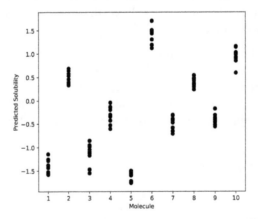

Figure 10-7. Solubilities of 10 molecules from the test set, as predicted by a set of models all trained on the same data.

This is a fundamentally different type of uncertainty, known as *epistemic uncertainty*. It comes from the fact that many different models fit the training data equally well, and we don't know which one is "best."

A straightforward way to measure epistemic uncertainty is to train many models and compare their results, as we have done in Figure 10-7. Often this is prohibitively expensive, however. If you have a large, complicated model that takes weeks to train, you don't want to repeat the process many times.

A much faster alternative is to train a single model using dropout, then predict each output many times with different dropout masks. Usually dropout is only performed at training time. If 50% of the outputs from a layer are randomly set to 0 in each training step, at test time you would instead multiply *every* output by 0.5. But let's not do that. Let's randomly set half the outputs to 0, then repeat the process with many different random masks to get a collection of different predictions. The variation between the predicted values gives a pretty good estimate of the epistemic uncertainty.

Notice how your modeling choices involve trade offs between these two kinds of uncertainty. If you use a large model with lots of parameters, you can get it to fit the training data very closely. That model will probably be underdetermined, however, so lots of combinations of parameter values will fit the training data equally well. If instead you use a small model with few parameters, there is more likely to be a unique set of optimal parameter values, but it probably also won't fit the training set as well. In either case, both types of uncertainty must be included when estimating the accuracy of the model's predictions.

This sounds complicated. How do we do it in practice? Fortunately, DeepChem makes it very easy. Just include one extra argument to the model's constructor:

```
model = GraphConvModel(n_tasks=1, mode='regression',
                       dropout=0.2, uncertainty=True)
```

By including the option `uncertainty=True`, we tell the model to add the extra outputs for uncertainty and make necessary changes to the loss function. Now we can make predictions like this:

```
y_pred, y_std = model.predict_uncertainty(test_dataset)
```

This computes the model's output many times with different dropout masks, then returns the average value for each output element, along with an estimate of the standard deviation of each one.

Figure 10-8 shows how it works on the test set. For each sample, we plot the actual error in the prediction versus the model's uncertainty estimate. The data shows a clear trend: samples with large predicted uncertainty tend to have larger errors than those with small predicted uncertainty. The dotted line corresponds to $y = 2x$. Points below this line have predicted solubilities that are within two (predicted) standard deviations of the true value. Roughly 90% of the samples are within this region.

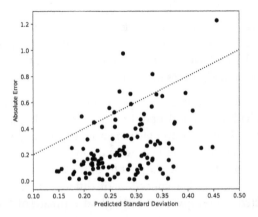

Figure 10-8. True error in the model's predictions, versus its estimates of the uncertainty in each value.

Interpretability, Explainability, and Real-World Consequences

The greater the consequences of a wrong prediction, the more important it is to understand how the model works. For some models, individual predictions are unimportant. A chemist working in the early stages of drug development might use a model to screen millions of potential compounds and select the most promising ones to synthesize. The accuracy of the model's predictions may be low, but that is acceptable. As long as the passing compounds are, on average, better than the rejected ones, it is serving a useful purpose.

In other cases, every prediction matters. When a model is used to diagnose a disease or recommend a treatment, the accuracy of each result can literally determine whether a patient lives or dies. The question "Should I trust this result?" becomes vitally important.

Ideally the model should produce not just a diagnosis, but also a summary of the evidence supporting that diagnosis. The patient's doctor could examine the evidence and make an informed decision about whether the model has functioned correctly in that particular case. A model that has this property is said to be *explainable*.

Unfortunately, far too many deep learning models are not explainable. In that case, the doctor is faced with a difficult choice. Do they trust the model, even if they have no idea what evidence a result is based on? Or do they ignore the model and rely on their own judgment? Neither choice is satisfactory.

Remember this principle: *every model ultimately interacts with humans.* To evaluate the quality of a model, you must include those interactions in your analysis. Often they depend as much on psychology or economics as on machine learning. It is not enough to compute a correlation coefficient or ROC AUC on the model's predictions. You must also consider who will see those predictions, how they will be interpreted, and what real-world effects they will ultimately have.

Making a model more interpretable or explainable may not affect the accuracy of its predictions, but it can still have a huge impact on the real-world consequences of those predictions. It is an essential part of model design.

Conclusion

Deep models have a reputation of being hard to interpret, but many useful techniques have been developed that can help. By using these techniques you can begin to understand what your model is doing and how it is working. That helps you decide whether to trust it, and lets you identify situations where it is likely to fail. It also may give new insights into the data. For example, by analyzing the TF binding model we discovered the binding motif for a particular transcription factor.

## CHAPTER 11

A Virtual Screening Workflow Example

Virtual screening can provide an efficient and cost-effective means of identifying starting points for drug discovery programs. Rather than carrying out an expensive, experimental high-throughput screen (HTS), we can use computational methods to virtually evaluate millions, or even tens of millions, of molecules. Virtual screening methods are often grouped into two categories, structure-based virtual screening and ligand-based virtual screening.

In a structure-based virtual screen, computational methods are used to identify molecules that will optimally fit into a cavity, known as a binding site, in a protein. The binding of a molecule into the protein binding site can often inhibit the function of the protein. For instance, proteins known as enzymes catalyze a variety of physiological chemical reactions. By identifying and optimizing inhibitors of these enzymatic processes, scientists have been able to develop treatments for a wide range of diseases in oncology, inflammation, infection, and other therapeutic areas.

In a ligand-based virtual screen, we search for molecules that function similarly to one or more known molecules. We may be looking to improve the function of an existing molecule, to avoid pharmacological liabilities associated with a known molecule, or to develop novel intellectual property. A ligand-based virtual screen typically starts with a set of known molecules identified through any of a variety of experimental methods. Computational methods are then used to develop a model based on experimental data, and this model is used to virtually screen a large set of molecules to find new chemical starting points.

In this chapter, we will walk through a practical example of a virtual screening workflow. We will examine the code used to carry out components of the virtual screen as well as the thought process behind decisions made throughout the analysis. In this particular case, we will carry out a ligand-based virtual screen. We will use a set of molecules known to bind to a particular protein, as well as a set of molecules assumed to not bind, to train a convolutional neural network to identify new molecules with the potential to bind to the target.

Preparing a Dataset for Predictive Modeling

As a first step, we will build a graph convolution model to predict the ability of molecules to inhibit a protein known as ERK2. This protein, also known as mitogen-activated protein kinase 1, or MAPK1, plays an important role in the signaling pathways that regulate how cells multiply. ERK2 has been implicated in a number of cancers, and ERK2 inhibitors are currently being tested in clinical trials for non-small-cell lung cancer and melanoma (skin cancer).

We will train the model to distinguish a set of ERK2 active compounds from a set of decoy compounds. The active and decoy compounds are derived from the DUD-E database (*http://dud.docking.org/*), which is designed for testing predictive models. In practice, we would typically obtain active and inactive molecules from the scientific literature, or from a database of biologically active molecules such as the ChEMBL database (*https://www.ebi.ac.uk/chembl/*) from the European Bioinformatics Institute (EBI). In order to generate the best model, we would like to have decoys with property distributions similar to those of our active compounds. Let's say this was not the case and the inactive compounds had lower molecular weight than the active compounds. In this case, our classifier might be trained simply to separate low molecular weight compounds from high molecular weight compounds. Such a classifier would have very limited utility in practice.

In order to better understand the dataset, let's examine a few calculated properties of our active and decoy molecules. To build a reliable model, we need to ensure that the properties of the active molecules are similar to those of the decoy molecules.

First, let's import the necessary libraries:

```
from rdkit import Chem           # RDKit libraries for chemistry functions
from rdkit.Chem import Draw      # Drawing chemical structures
import pandas as pd              # Dealing with data in tables
from rdkit.Chem import PandasTools # Manipulating chemical data
from rdkit.Chem import Descriptors # Calculating molecular descriptors
from rdkit.Chem import rdmolops  # Additional molecular properties
import seaborn as sns            # Making graphs
```

In this exercise, molecules are represented using SMILES strings. For more information on SMILES, please see Chapter 4. We can now read a SMILES file into a Pandas dataframe and add an RDKit molecule to the dataframe. While the input SMILES file is not technically a CSV file, the Pandas read_CSV() function can read it as long as we specify the delimiter, which in this case is a space:

```
active_df = pd.read_CSV("mk01/actives_final.ism",header=None,sep=" ")
active_rows,active_cols = active_df.shape
```

```
active_df.columns = ["SMILES","ID","ChEMBL_ID"]
active_df["label"] = ["Active"]*active_rows
PandasTools.AddMoleculeColumnToFrame(active_df,"SMILES","Mol")
```

Let's define a function to add the calculated properties to a dataframe:

```
def add_property_columns_to_df(df_in):
df_in["mw"] = [Descriptors.MolWt(mol) for mol in
df_in.Mol]
df_in["logP"] = [Descriptors.MolLogP(mol) for mol in
df_in.Mol]
df_in["charge"] = [rdmolops.GetFormalCharge(mol) for mol
in df_in.Mol]
```

With this function in hand, we can calculate the molecular weight, LogP, and formal charge of the molecules. These properties encode the size of a molecule, its ability to partition from an oily substance (octanol) to water, and whether the molecule has a positive or negative charge. Once we have these properties we can compare the distributions for the active and decoy sets:

```
add_property_columns_to_df(active_df)
```

Let's look at the first few rows of our dataframe to ensure that the contents of the dataframe match the input file (see Table 11-1):

```
active_df.head()
```

Table 11-1. The first few lines of the active_df dataframe.

	SMILES	ID	ChEMBL_ID	label
0	Cn1ccnc1Sc2ccc(cc2Cl)Nc3c4cc(c(cc4ncc3C#N)OCCCN5CCOCC5)OC	168691	CHEMBL318804	Active
1	C[C@@]12[C@@H]([C@@H](CC(O1)n3c4ccccc4c5c3c6n2c7ccccc7c6c8c5C(=O)NC8)NC)OC	86358	CHEMBL162	Active
2	Cc1cnc(nc1c2cc([nH]c2)C(=O)N[C@H](CO)c3cccc(c3)Cl)Nc4cccc5c40C(O5)(F)F	575087	CHEMBL576683	Active
3	Cc1cnc(nc1c2cc([nH]c2)C(=O)N[C@H](CO)c3cccc(c3)Cl)Nc4cccc5c40C05	575065	CHEMBL571484	Active
4	Cc1cnc(nc1c2cc([nH]c2)C(=O)N[C@H](CO)c3cccc(c3)Cl)Nc4cccc5c4CCC5	575047	CHEMBL568937	Active

Now let's do the same thing with the decoy molecules:

```
decoy_df = pd.read_CSV("mk01/decoys_final.ism",header=None,sep=" ")
decoy_df.columns = ["SMILES","ID"]
decoy_rows, decoy_cols = decoy_df.shape
decoy_df["label"] = ["Decoy"]*decoy_rows
PandasTools.AddMoleculeColumnToFrame(decoy_df,"SMILES","Mol")
add_property_columns_to_df(decoy_df)
```

In order to build a model, we need a single dataframe with the active and decoy molecules. We can use the Pandas append function to add the two dataframes and create a new dataframe called tmp_df:

```
tmp_df = active_df.append(decoy_df)
```

With properties calculated for both the active and decoy sets, we can compare the properties of the two sets of molecules. To do the comparison, we will use violin plots. A violin plot is analogous to a boxplot. The violin plot provides a mirrored, horizontal view of a frequency distribution. Ideally, we would like to see similar distributions for the active and decoy sets. The results are shown in Figure 11-1:

```
sns.violinplot(tmp_df["label"],tmp_df["mw"])
```

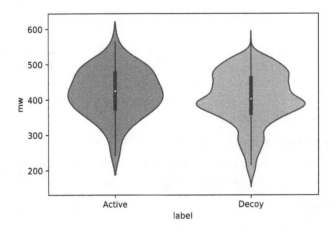

Figure 11-1. Violin plots of molecular weight for the active and decoy sets.

An examination of these plots shows that the molecular weight distributions for the two sets are roughly equivalent. The decoy set has more low molecular weight molecules, but the center of the distribution, shown as a box in the middle of each violin plot, is in a similar location in both plots.

We can use violin plots to perform a similar comparison of the LogP distributions (Figure 11-2). Again, we can see that the distributions are similar, with a few more of the decoy molecules at the lower end of the distribution:

```
sns.violinplot(tmp_df["label"],tmp_df["logP"])
```

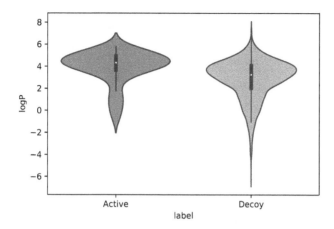

Figure 11-2. Violin plots of LogP for the active and decoy sets.

Finally, we perform the same comparison with the formal charges of the molecules (Figure 11-3):

```
sns.violinplot(new_tmp_df["label"],new_tmp_df["charge"])
```

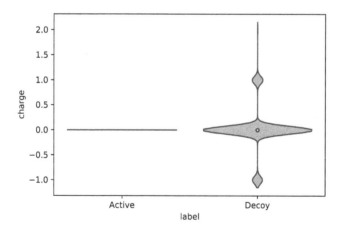

Figure 11-3. Violin plots of formal charge for the active and decoy sets.

In this case, we see a significant difference. All of the active molecules are neutral, having charges of 0, while some of the decoys are charged, with charges of +1 or −1. Let see what fraction of the decoy molecules are charged. We can do this by creating a new dataframe with just the charged molecules:

```
charged = decoy_df[decoy_df["charge"] != 0]
```

A Pandas dataframe has a property, `shape`, that returns the number of rows and columns in the dataframe. As such, element [0] in the `shape` property will be the number of rows. Let's divide the number of rows in our dataframe of charged molecules by the total number of rows in the decoy dataframe:

```
charged.shape[0]/decoy_df.shape[0]
```

This returns 0.162. As we saw in the violin plot, approximately 16% of the decoy molecules are charged. This appears to be because the active and decoy sets were not prepared in a consistent fashion. We can address this problem by modifying the chemical structures of the decoy molecules to neutralize their charges. Fortunately we can do this easily with the `NeutraliseCharges()` function from the RDKit Cookbook (*https://www.rdkit.org/docs/Cookbook.html*):

```
from neutralize import NeutraliseCharges
```

In order to avoid confusion, we create a new dataframe with the SMILES stings, IDs, and labels for the decoys:

```
revised_decoy_df = decoy_df[["SMILES","ID","label"]].copy()
```

With this new dataframe in hand, we can replace the original SMILES strings with the strings for the neutral forms of the molecules. The `NeutraliseCharges` function returns two values. The first is the SMILES string for the neutral form of the molecule and the second is a Boolean variable indicating whether the molecule was changed. In the following code, we only need the SMILES string, so we use the first element of the tuple returned by `NeutraliseCharges`.

```
revised_decoy_df["SMILES"] = [NeutraliseCharges(x)[0] for x
in revised_decoy_df["SMILES"]]
```

Once we've replaced the SMILES strings, we can add a molecule column to our new dataframe and calculate the properties again:

```
PandasTools.AddMoleculeColumnToFrame(revised_decoy_df,"SMILES","Mol")
add_property_columns_to_df(revised_decoy_df)
```

We can then append the dataframe with the active molecules to the one with the revised, neutral decoys:

```
new_tmp_df = active_df.append(revised_decoy_df)
```

Next, we can generate a new boxplot to compare the charge distributions of the active molecules with those of our neutralized decoys (Figure 11-4):

```
sns.violinplot(new_tmp_df["label"],new_tmp_df["charge"])
```

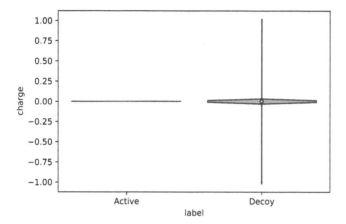

Figure 11-4. Violin plots of the charge distribution for our revised decoy set.

An examination of the plots shows that there are now very few charged molecules in the decoy set. We can use the same technique we used earlier to create a dataframe with only the charged molecules. We then use this dataframe to determine the number of charged molecules remaining in the set:

```
charged = revised_decoy_df[revised_decoy_df["charge"] != 0]
charged.shape[0]/revised_decoy_df.shape[0]
```

The result now is 0.003. We have reduced the fraction of charged molecules from 16% to 0.3%. We can now be confident that our active and decoy sets are reasonably well balanced.

In order to use these datasets with DeepChem, we need to write the molecules out as a CSV file containing for each molecule the SMILES string, ID, Name, and an integer value indicating whether the compounds are active (labeled as 1) or inactive (labeled as 0):

```
active_df["is_active"] = [1] * active_df.shape[0]
revised_decoy_df["is_active"] = [0] * revised_decoy_df.shape[0]
combined_df = active_df.append(revised_decoy_df)[["SMILES","ID","is_active"]]
combined_df.head()
```

The first five lines are shown in Table 11-2.

Table 11-2. The first few lines of our new combined dataframe

	SMILES	ID	is_active
0	Cn1 ccnc1Sc2ccc(cc2Cl}Nc3c4cc(c(cc4ncc3C#N}OCCCN5CCOCC5)OC	168691	1
1	C[C@@]12[C@@H]([C@@H](CC(O1)n3c4ccccc4c5c3c6n2c7ccccc7c6c8c5C(=O)NC8)NC)OC	86358	1
2	Cc1cnc(nc1c2cc([nH]c2)C(=O) N[C@H](CO)c3cccc(c3}Cl}Nc4cccc5c4OC(O5)(F)F	575087	1
3	CCc1cnc(nc1c2cc([nH]c2)C(=O)N[C@H](CO)c3cccc(c3}Cl}Nc4cccc5c4OC05	575065	1
4	Cc1cnc(nc1c2cc([nH]c2)C(=O) N[C@H](CO)c3cccc(c3}Cl}Nc4cccc5c4CCC5	575047	1

Our final step in this section is to save our new `combined_df` as a CSV file. The `index=False` option causes Pandas to not include the row number in the first column:

```
combined_df.to_csv("dude_erk1_mk01.CSV",index=False)
```

Training a Predictive Model

Now that we have taken care of formatting, we can use this data to train a graph convolution model. First, we need to import the necessary libraries. Some of these libraries were imported in the first section, but let's assume we are starting with the CSV file we created in the previous section:

```
import deepchem as dc                              # DeepChem libraries
from deepchem.models import GraphConvModel # Graph convolutions
import numpy as np                                 # NumPy for numeric operations
import sys                                         # Error handling
import pandas as pd                                # Data table manipulation
import seaborn as sns                              # Seaborn library for plotting
from rdkit.Chem import PandasTools                 # Chemical structures in Pandas
```

Now let's define a function to create a `GraphConvModel`. In this case, we will be creating a classification model. Since we will apply the model later on a different dataset, it's a good idea to create a directory in which to store the model. You will need to change the directory to something accessible on your filesystem:

```
def generate_graph_conv_model():
batch_size = 128
model = GraphConvModel(1, batch_size=batch_size,
mode='classification',
model_dir="/tmp/mk01/model_dir")
return model
```

In order to train the model, we first read in the CSV file we created in the previous section:

```
dataset_file = "dude_erk2_mk01.CSV"
tasks = ["is_active"]
featurizer = dc.feat.ConvMolFeaturizer()
```

```
loader = dc.data.CSVLoader(tasks=tasks,
smiles_field="SMILES",
featurizer=featurizer)
dataset = loader.featurize(dataset_file, shard_size=8192)
```

Now that we have the dataset loaded, let's build a model. We will create training and test sets to evaluate the model's performance. In this case, we will use the `Random Splitter` (DeepChem offers a number of other splitters too, such as the `ScaffoldS plitter`, which divides the dataset by chemical scaffold, and the `ButinaSplitter`, which first clusters the data then splits the dataset so that different clusters end up in the training and test sets):

```
splitter = dc.splits.RandomSplitter()
```

With the dataset split, we can train a model on the training set and test that model on the validation set. At this point, we need to define some metrics and evaluate the performance of our model. In this case, our dataset is unbalanced: we have a small number of active compounds and a large number of inactive compounds. Given this difference, we need to use a metric that reflects the performance on unbalanced datasets. One metric that is appropriate for datasets like this is the Matthews correlation coefficient (MCC):

```
metrics = [
dc.metrics.Metric(dc.metrics.matthews_corrcoef, np.mean,
mode="classification")]
```

In order to evaluate the performance of our model, we will perform 10 folds of cross validation, where we train a model on the training set and validate on the validation set:

```
training_score_list = []
validation_score_list = []
transformers = []
cv_folds = 10
for i in range(0, cv_folds):
model = generate_graph_conv_model()
res = splitter.train_valid_test_split(dataset)
train_dataset, valid_dataset, test_dataset = res
model.fit(train_dataset)
train_scores = model.evaluate(train_dataset, metrics,
transformers)
training_score_list.append(
train_scores["mean-matthews_corrcoef"])
validation_scores = model.evaluate(valid_dataset,
metrics,
transformers)
validation_score_list.append(
validation_scores["mean-matthews_corrcoef"])
print(training_score_list)
print(validation_score_list)
```

To visualize the performance of our model on the training and test data, we can make boxplots. The results are shown in Figure 11-5:

```
sns.boxplot(
["training"] * cv_folds + ["validation"] * cv_folds,
training_score_list + validation_score_list)
```

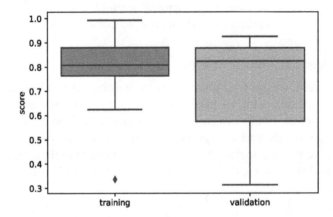

Figure 11-5. Boxplots of scores for the training and validation sets.

The plots indicate that, as expected, the performance on the training set is superior to that on the validation set. However, the performance on the validation set is still quite good. At this point, we can be confident in the performance of our model.

It is also useful to visualize the results of our model. In order to do this, we will generate a set of predictions for a validation set:

```
pred = [x.flatten() for x in model.predict(valid_dataset)]
```

To make processing easier, we'll create a Pandas dataframe with the predictions:

```
pred_df = pd.DataFrame(pred,columns=["neg","pos"])
```

We can easily add the activity class (1 = active, 0 = inactive) and the SMILES strings for our predicted molecules to the dataframe:

```
pred_df["active"] = [int(x) for x in valid_dataset.y]
pred_df["SMILES"] = valid_dataset.ids
```

It's always a good idea to look at the first few lines of the dataframe to ensure that the data makes sense. Table 11-3 shows the results.

Table 11-3. The first few lines of the dataframe containing the predictions

	neg	pos	active	SMILES
0	0.906081	0.093919	1	Cn1ccnc1Sc2ccc(cc2Cl)Nc3c4cc(c(cc4ncc3C#N)OCCC...
1	0.042446	0.957554	1	Cc1cnc(nc1c2cc([nH]c2)C(=O)N[C@H](CO)c3cccc(c3...
2	0.134508	0.865492	1	Cc1cccc(c1)[C@@H](CO)NC(=O)c2cc(c[nH]2)c3c(cnc...
3	0.036508	0.963492	1	Cc1cnc(nc1c2cc([nH]c2)C(=O)N[C@H](CO)c3ccccc3)...
4	0.940717	0.059283	1	c1c\2c([nH]c1Br)C(=O)NCC/C2=C/3\C(=O)N=C(N3)N

Creating boxplots enables us to compare the predicted values for the active and inactive molecules (see Figure 11-6).

```
sns.boxplot(pred_df.active,pred_df.pos)
```

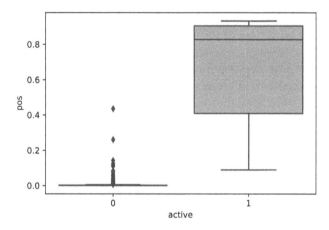

Figure 11-6. Positive scores for the predicted molecules.

The performance of our model is very good: we can see a clear separation between the active and inactive molecules. When building a predictive model it is often important to examine inactive molecules that are predicted as active (false positives) as well as active molecules that are predicted as inactive (false negatives). It appears that only one of our active molecules received a low positive score. In order to look more closely, we will create a new dataframe containing all of the active molecules with a positive score < 0.5):

```
false_negative_df = pred_df.query("active == 1 & pos < 0.5").copy()
```

To inspect the chemical structures of the molecules in our dataframe, we use the Pan dasTools module from RDKit:

```
PandasTools.AddMoleculeColumnToFrame(false_negative_df,
    "SMILES", "Mol")
```

Let's look at the new dataframe (Figure 11-7):

```
false_negative_df
```

	neg	pos	active	SMILES	Mol
4	0.723421	0.27658	1	c1ccc(cc1)c2c(c3ccccn3n2)c4cc5c(n[nH]c5nn4)N	
5	0.910040	0.08996	1	CCNC(=O)Nc1ccc(cn1)CNc2c(scn2)C(=O)Nc3ccc4c(c3)OC(O4)(F)F	

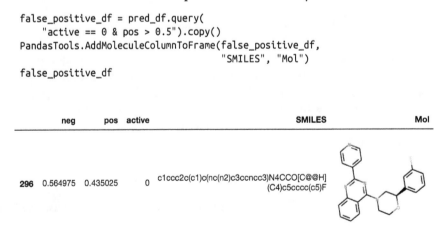

Figure 11-7. False negative predictions.

In order to fully take advantage of the information in this dataframe, we need to have some knowledge of medicinal chemistry. It is often informative to look at the chemical structures of the false negative molecules and compare these with the chemical structures of the true positive molecules. This may provide some insight into the reasons that molecules were not predicted correctly. Often it may be the case that the false negative molecules are not similar to any of the true positive molecules. In this case, it may be worth performing additional literature searches to increase the diversity of the molecules in the training set.

We can use a similar approach to examine the false positive molecules, which are inactive but received a positive score > 0.5 (see Figure 11-8). Again, comparison with the chemical structures of the true positive molecules may be informative:

```
false_positive_df = pred_df.query(
    "active == 0 & pos > 0.5").copy()
PandasTools.AddMoleculeColumnToFrame(false_positive_df,
                        "SMILES", "Mol")

false_positive_df
```

	neg	pos	active	SMILES	Mol
296	0.564975	0.435025	0	c1ccc2c(c1)c(nc(n2)c3ccncc3)N4CCO[C@@H](C4)c5cccc(c5)F	

Figure 11-8. A false positive molecule.

During the model training phase, our objective was to evaluate the performance of our model. As such, we trained the model on a portion of the data and validated the model on the remainder. Now that we have evaluated the performance, we want to generate the most effective model. In order to do this, we will train the model on all of the data:

```
model.fit(dataset)
```

This gives us an accuracy score of 91%. Finally, we save the model to disk so that we can use it to make future predictions:

```
model.save()
```

Preparing a Dataset for Model Prediction

Now that we've created a predictive model, we can apply this model to a new set of molecules. In many cases, we will build a predictive model based on literature data, then apply that model to a set of molecules that we want to screen. The molecules we want to screen may come from an internal database or from a commercially available screening collection. As an example, we will use the predictive model we created to screen a small sample of 100,000 compounds from the ZINC database, a collection of more than 1 billion commercially available molecules.

One potential source of difficulty when carrying out a virtual screen is the presence of molecules that have the potential to interfere with biological assays. Over the last 25 years, many groups within the scientific community have developed sets of rules to identify potentially reactive or problematic molecules. Several of these rule sets, which are encoded as SMARTS strings, have been collected by the group that curates the ChEMBL database. These rule sets have been made available through a Python script called *rd_filters.py*. In this example, we will use *rd_filters.py* to identify potentially problematic molecules in our set of 100,000 molecules from the ZINC database.

The *rd_filters.py* script and associated data files are available on our GitHub repository (*https://github.com/deepchem/DeepLearningLifeSciences*).

The available modes and arguments for the script can be obtained by calling it with the -h flag.

```
rd_filters.py -h

Usage:
rd_filters.py $ filter --in INPUT_FILE --prefix PREFIX [--rules RULES_FILE_NAME]
[--alerts ALERT_FILE_NAME][--np NUM_CORES]
rd_filters.py $ template --out TEMPLATE_FILE [--rules RULES_FILE_NAME]
Options:
--in INPUT_FILE input file name
--prefix PREFIX prefix for output file names
--rules RULES_FILE_NAME name of the rules JSON file
--alerts ALERTS_FILE_NAME name of the structural alerts file
```

```
--np NUM_CORES the number of cpu cores to use (default is all)
--out TEMPLATE_FILE parameter template file name
```

To call the script on our input file, which is called *zinc_100k.smi*, we can specify the input file and a prefix for output filenames. The filter argument calls the script in "filter" mode, where it identifies potentially problematic molecules. The --prefix argument indicates that the output file names will start with the prefix *zinc*.

```
rd_filters.py filter --in zinc_100k.smi --prefix zinc

using 24 cores
Using alerts from Inpharmatica
Wrote SMILES for molecules passing filters to zinc.smi
Wrote detailed data to zinc.CSV
68752 of 100000 passed filters 68.8%
Elapsed time 15.89 seconds
```

The output indicates the following:

- The script is running on 24 cores. It runs in parallel across multiple cores, and the number of cores can be selected with the -np flag.

- The script is using the "Inpharmatica" set of rules. This rule set covers a large range of chemical functionality that has been shown to be problematic in biological assays. In addition to the Inpharmaticia set, the script has seven other rule sets available. Please see the *rd_filters.py* documentation for more information.

- SMILES strings for the molecules passing the filters were written to a file called *zinc.smi*. We will use this as the input when we use the predictive model.

- Detailed information on which compounds triggered particular structural alerts was written to a file called *zinc.CSV*.

- 69% of the molecules passed the filters, and 31% were considered problematic.

It is informative to take a look at the reasons why 31% of the molecules were rejected. This can let us know whether we need to adjust any of the filters. We will use a bit of Python code to look at the first few lines of output, shown in Table 11-4.

```
import pandas as pd
df = pd.read_CSV("zinc.CSV")
df.head()
```

Table 11-4. The first few lines of the dataframe created from zinc.CSV

	SMILES	NAME	FILTER	MW	LogP	HBD
0	CN(CCO)C[C@@H](O)Cn1cnc2c1c(=O)n(C)c(=O)n2C	ZINC000000000843	Filter82_pyridinium >0	311.342	−2.2813	2
1	O=c1[nH]c(=O)n([C@@H]2C[C@@H](O)[C@H](CO)O2)cc1Br	ZINC000000001063	Filter82_pyridinium >0	307.100	−1.0602	3

	SMILES	NAME	FILTER	MW	LogP	HBD
2	Cn1c2ncn(CC(=O)N3CCOCC3)c2c(=O)n(C)c1=O	ZINC000000003942	Filter82_pyridinium >0	307.310	−1.7075	0
3	CN1C(=O)C[C@H] (N2CCN(C(=O)CN3CCCC3)CC2)C1=O	ZINC000000036436	OK	308.382	−1.0163	0
4	CC(=O)NC[C@H](O)[C@H]1O[C@H]2OC(C) (C)O[C@H]2[C...	ZINC 000000041101	OK	302.327	-1.1355	3

The dataframe has six columns:

SMILES
> the SMILES strings for each molecule.

NAME
> the molecule NAME, as listed in the input file.

FILTER
> the reason the molecule was rejected, or "OK" if the molecule was not rejected.

MW
> the molecular weight of the molecule. By default, molecules with molecular weight greater than 500 are rejected.

LogP
> the calculated octanol/water partition coefficient of the molecule. By default, molecules with LogP greater than five are rejected.

HBD
> the number of hydrogen bond donors. By default, molecules with more than 5 hydrogen bond donors are rejected.

We can use the Counter class from the Python collections library to identify which filters were responsible for removing the largest numbers of molecules (see Table 11-5):

```
from collections import Counter
count_list = list(Counter(df.FILTER).items())
count_df = pd.DataFrame(count_list,columns=["Rule","Count"])
count_df.sort_values("Count",inplace=True,ascending=False)
count_df.head()
```

Table 11-5. Counts of the number of molecules selected by the top 5 filters

	Rule	Count
1	OK	69156
6	Filter41_12_dicarbonyl > 0	19330
0	Filter82_pyridinium > 0	7761
10	Filter93_acetyl_urea > 0	1541
11	Filter78_bicyclic_Imide > 0	825

The first line in the table, labeled as "OK," indicates the number of molecules that were not eliminated by any of the filters. From this, we can see that 69,156 of the molecules in our input passed all of the filters. The largest number of molecules (19,330) were rejected because they contained a 1,2-dicarbonyl group. Molecules of this type may react and form covalent bonds with protein residues such as serine and cysteine. We can find the SMARTS pattern used to identify these molecules by looking for the string "Filter41_12_dicarbonyl" in the *filter_collection.CSV* file that is part of the *rd_filters.py* distribution. The SMARTS pattern is "*C(=O)C(=O)*", which represents:

- Any atom, connected to
- carbon double bonded to oxygen, connected to
- carbon double bonded to oxygen, connected to
- any atom.

It is always good to look at the data and ensure that everything is working as expected. We can use the highlightAtomLists argument to RDKit's MolsToGrid Image() function to highlight the 1,2-dicarbonyl functionality (see Figure 11-9):

```
from rdkit import Chem
from rdkit.Chem import Draw

mol_list = [Chem.MolFromSmiles(x) for x in smiles_list]
dicarbonyl = Chem.MolFromSmarts('*C(=O)C(=O)*')
match_list = [mol.GetSubstructMatch(dicarbonyl) for mol in
              mol_list]
Draw.MolsToGridImage(mol_list,
                     highlightAtomLists=match_list,
                     molsPerRow=5)
```

We can see that the molecules do indeed have dicarbonyl groups, as highlighted in the figure. If we wanted to, we could similarly evaluate other filters. At this point, however, we can be satisfied with the results of the filtering. We have removed the problematic molecules from the set we plan to use for our virtual screen. We can now use this set, which is in the file *zinc.smi*, in the next step of this exercise.

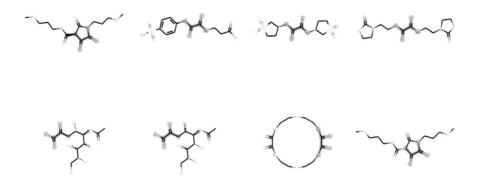

Figure 11-9. Molecules containing a 1,2-dicarbonyl group.

Applying a Predictive Model

The GraphConvMdel we created can now be used to search the set of commercially available compounds we just filtered. Applying the model requires a few steps:

1. Load the model from disk.

2. Create a featurizer.

3. Read and featurize the molecules that will run through the model.

4. Examine the scores for the predictions.

5. Examine the chemical structures of the top predicted molecules.

6. Cluster the selected molecules.

7. Write the selected molecules from each cluster to a CSV file.

We begin by importing the necessary libraries:

```
import deepchem as dc                            # DeepChem libraries
import pandas as pd                              # Pandas for tables
from rdkit.Chem import PandasTools, Draw         # Chemistry in Pandas
from rdkit import DataStructs                    # For fingerprint handling
from rdkit.ML.Cluster import Butina              # Cluster molecules
from rdkit.Chem import rdMolDescriptors as rdmd  # Descriptors
import seaborn as sns                            # Plotting
```

and loading the model we generated earlier:

```
model = dc.models.TensorGraph.load_from_dir(""/tmp/mk01/model_dir"")
```

To generate predictions from our model, we first need to featurize the molecules we plan to use to generate predictions. We do this by instantiating a DeepChem ConvMol Featurizer:

```
featurizer = dc.feat.ConvMolFeaturizer()
```

In order to featurize the molecules, we need to transform our SMILES file into a CSV
file. In order to create a DeepChem featurizer we also require an activity column, so
we add one, then write the file to CSV:

```
df = pd.read_CSV("zinc.smi",sep=" ",header=None)
df.columns=["SMILES","Name"]
rows,cols = df.shape
# Just add add a dummy column to keep the featurizer happy
df["Val"] = [0] * rows
```

As before, we should look at the first few lines of the file (shown in Table 11-6) to
make sure everything is as we had expect:

```
df.head()
```

Table 11-6. The first few lines of the input file

	SMILES	Name	Val
0	CN1C(=O)C[C@H](N2CCN(C(=O)CN3CCCC3)CC2)C1=O	ZINC000000036436	0
1	CC(=O)NC[C@H](O)[C@H]1O[C@H]2OC(C)(C)O[C@H]2[C@@H]1NC(C)=O	ZINC000000041101	0
2	C1CN(c2nc(-c3nn[nH]n3)nc(N3CCOCC3)n2)CCO1	ZINC000000054542	0
3	OCCN(CCO)c1nc(Cl)nc(N(CCO)CCO)n1	ZINC000000109481	0
4	COC(=O)c1ccc(S(=O)(=O)N(CCO)CCO)n1C	ZINC000000119782	0

Note that the Val column is just a placeholder to keep the DeepChem featurizer
happy. The file looks good, so we will write it as a CSV file to use as input for Deep-
Chem. We use the index=False argument to prevent Pandas from writing the row
numbers as the first column:

```
infile_name = "zinc_filtered.CSV"
df.to_CSV(infile_name,index=False)
```

We can use DeepChem to read this CSV file with a loader and featurize the molecules
we plan to predict:

```
loader = dc.data.CSVLoader(tasks=['Val'],
                           smiles_field="SMILES",
                           featurizer=featurizer)
dataset = loader.featurize(infile_name, shard_size=8192)
```

The featurized molecules can be used to generate predictions with the model:

```
pred = model.predict(dataset)
```

For convenience,we will put the predictions into a Pandas dataframe:

```
pred_df = pd.DataFrame([x.flatten() for x in pred],
columns=["Neg", "Pos"])
```

The distribution plot, available in the Seaborn library, provides a nice overview of the distribution of scores. Unfortunately, in virtual screening, there are no clear rules for defining an activity cutoff. Often the best strategy is to look at the distribution of scores, then select a set of the top-scoring molecules. If we look at the plot in Figure 11-10, we can see that there are only a small number of molecules with scores above 0.3. We can use this value as a preliminary cutoff for molecules that we may want to screen experimentally.

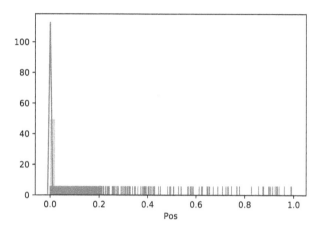

Figure 11-10. Distribution plot of the scores for the predicted molecules.

We can join the dataframe with the scores to the dataframe with the SMILES strings. This gives us the ability to view the chemical structures of the top-scoring molecules:

```
combo_df = df.join(pred_df, how="outer")
combo_df.sort_values("Pos", inplace=True, ascending=False)
```

As we saw earlier, adding a molecule column to the dataframe enables us to look at the chemical structures of the hits (see Figure 11-11).

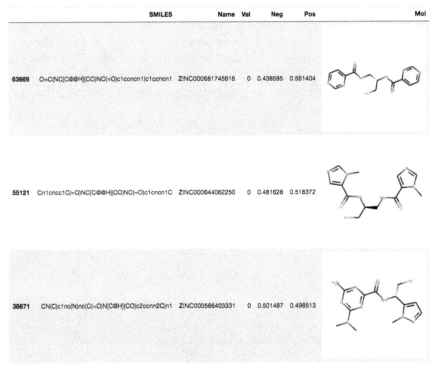

	SMILES	Name	Val	Neg	Pos	Mol
63669	O=C(NC[C@@H](CO)NC(=O)c1ccncn1)c1ccncn1	ZINC000681745616	0	0.438595	0.561404	
55121	Cn1cncc1C(=O)NC[C@@H](CO)NC(=O)c1cncn1C	ZINC000644062250	0	0.481628	0.518372	
38671	CN(C)c1nc(N)nc(C(=O)N[C@H](CO)c2ccnn2C)n1	ZINC000566403331	0	0.501487	0.498513	

Figure 11-11. Chemical structures of the top-scoring molecules.

Based on what we see here, it looks like many of the hits are similar. Let's look at a few more molecules (Figure 11-12):

```
Draw.MolsToGridImage(combo_df.Mol[:10], molsPerRow=5,
                     legends=["%.2f" % x for x in combo_df.Pos[:10]])
```

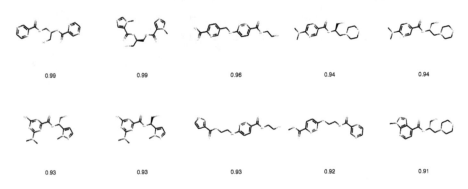

Figure 11-12. Structure grid with top-scoring hits. Values below the structures are model scores.

Indeed, many of the molecules are very similar and might end up being redundant in our screen. One way to be more efficient would be to cluster the molecules and only screen the highest-scoring molecule in each cluster. RDKit has an implementation of the Butina clustering method, one of the most highly used methods in cheminformatics. In the Butina clustering method, we group molecules based on their chemical similarity, which is calculated using a comparison of bit vectors (arrays of 1 and 0), also known as *chemical fingerprints* that represent the presence or absence of patterns of connected atoms in a molecule. These bit vectors are typically compared using a metric known as the *Tanimoto coefficient*, which is defined as:

$$Tanimoto = \frac{A \cap B}{A \cup B}$$

The numerator of the equation is the intersection, or the number of bits that are 1 in both bit vectors A and B. The denominator is the number of bits that are 1 in either vector A or vector B. The Tanimoto coefficient can range between 0, indicating that the molecules have no patterns of atoms in common, and 1, indicating that all of the patterns contained in molecule A are also contained in molecule B. As an example, we can consider the bit vectors shown in Figure 11-13. The intersection of the two vectors is 3 bits, while the union is 5. The Tanimoto coefficient is then 3/5, or 0.6. Note that the example shown here has been simplified for demonstration purposes. In practice, these bit vectors can contain hundreds or even thousands of bits.

```
A   =    11011010
B   =    11010000
A∩B =    11010000   Intersection = 3

A   =    11011010
B   =    11010000
A∪B =    11011010   Union = 5
```

Figure 11-13. Calculating a Tanimoto coefficient.

A small amount of code is necessary to cluster a set of molecules. The only parameter required for Butina clustering is the cluster cutoff. If the Tanimoto similarity of two molecules is greater than the cutoff, the molecules are put into the same cluster. If the similarity is less than the cutoff, the molecules are put into different clusters:

```python
def butina_cluster(mol_list, cutoff=0.35):
    fp_list = [
        rdmd.GetMorganFingerprintAsBitVect(m, 3, nBits=2048)
        for m in mol_list]
    dists = []
    nfps = len(fp_list)
```

```
for i in range(1, nfps):
    sims = DataStructs.BulkTanimotoSimilarity(
        fp_list[i], fp_list[:i])
    dists.extend([1 - x for x in sims])
mol_clusters = Butina.ClusterData(dists, nfps, cutoff,
                                  isDistData=True)
cluster_id_list = [0] * nfps
for idx, cluster in enumerate(mol_clusters, 1):
    for member in cluster:
        cluster_id_list[member] = idx
return cluster_id_list
```

Before clustering, we will create a new dataframe with only the 100 top-scoring molecules. Since `combo_df` is already sorted, we only have to use the `head` function to select the first 100 rows in the dataframe:

```
best_100_df = combo_df.head(100).copy()
```

We can then create a new column containing the cluster identifier for each compound:

```
best_100_df["Cluster"] = butina_cluster(best_100_df.Mol)
best_100_df.head()
```

As always, it's good to take a look and make sure everything worked. We now see that in addition to the SMILES string, molecule name, and predicted values, we also have a cluster identifier (see Figure 11-14).

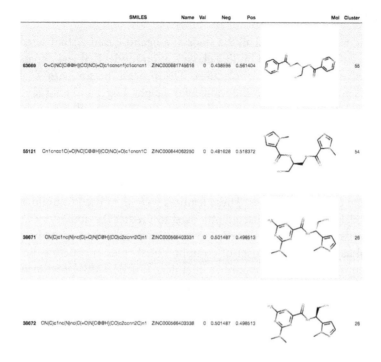

	SMILES	Name	Val	Neg	Pos		Mol	Cluster
63669	O=C(NC[C@@H](CO)NC(=O)c1ccncn1)c1ccncn1	ZINC000881745616	0	0.438596	0.561404			55
55121	Cn1cncc1C(=O)NC[C@@H](CO)NC(=O)c1cncn1C	ZINC000644062250	0	0.481628	0.518372			54
38671	CN(C)c1nc(N)nc(C(=O)N[C@H](CO)c2ccnn2C)n1	ZINC000566403331	0	0.501487	0.498513			26
38672	CN(C)c1nc(N)nc(C(=O)N[C@@H](CO)c2ccnn2C)n1	ZINC000566403338	0	0.501487	0.498513			26

Figure 11-14. The first few rows of the clustered dataset.

We can use the Pandas `unique` function to determine that we have 55 unique clusters:

```
len(best_100_df.Cluster.unique())
```

Ultimately, we would like to purchase these compounds and screen them experimentally. In order to do this, we need to save a CSV file listing the molecules we plan to purchase. The `drop_duplicates` function can be used to select one molecule per cluster. By default, the function starts from the top of the table and removes rows with values that have already been seen:

```
best_cluster_rep_df = best_100_df.drop_duplicates("Cluster")
```

Just to make sure that this operation worked, let's use the `shape` parameter to get the number of rows and columns in the new dataframe:

```
best_cluster_rep_df.shape
```

Finally, we can write out a CSV file with the molecules we want to screen:

```
best_cluster_rep_df.to_CSV("best_cluster_represenatives.CSV")
```

Conclusion

At this point, we have followed all the steps of a ligand-based virtual screening work-flow. We used deep learning to build a classification model that was capable of distinguishing active from inactive molecules. The process began with evaluating our training data and ensuring that the molecular weight, LogP, and charge distributions were balanced between the active and decoy sets. Once we'd made the necessary adjustments to the chemical structures of the decoy molecules, we were ready to build a model.

The first step in building the model was generating a set of chemical features for the molecules being used. We used the DeepChem GraphConv featurizer to generate a set of appropriate chemical features. These features were then used to build a graph convolution model, which was subsequently used to predict the activity of a set of commercially available molecules. In order to avoid molecules that could be problematic in biological assays, we used a set of computational rules encoded as SMARTS patterns to identify molecules containing chemical functionality previously known to interfere with assays or create subsequent liabilities.

With our list of desired molecules in hand, we are in a position to test these molecules in biological assays. Typically the next step in our workflow would be to obtain samples of the chemical compounds for testing. If the molecules came from a corporate compound collection, a robotic system would collect the samples and prepare them for testing. If the molecules were purchased from commercial sources, additional weighing and dilution with buffered water or another solvent would be necessary.

Once the samples are prepared, they are tested in biological assays. These assays can cover a wide range of endpoints, ranging from inhibiting bacterial growth to preventing the proliferation of cancer cells. While the testing of these molecules is the final step in our virtual screening exercise, it is far from the end of the road for a drug discovery project. Once we have run the initial biological assay on the molecules we identified through virtual screening, we analyze the results of the screen. If we find experimentally active molecules, we will typically identify and test other similar molecules that will enable us to understand the relationships between different parts of the molecule and the biological activity that we are measuring. This optimization process often involves the synthesis and testing of hundreds or even thousands of molecules to identify those with the desired combination of safety and biological activity.

## CHAPTER 12

Prospects and Perspectives

The life sciences are advancing at a remarkable rate, perhaps faster than any other branch of science. The same can be said of deep learning: it is one of the most exciting, rapidly advancing areas of computer science. The combination of the two has the potential to change the world in dramatic, far-reaching ways. The effects are already starting to be felt, but those are trivial compared to what will likely happen over the next few decades. The union of deep learning with biology can do enormous good, but also great harm.

In this final chapter we will set aside the mechanics of training deep models and take a broader view of the future of the field. Where does it have the greatest potential to solve important problems in the coming years? What obstacles must be overcome for that to happen? And what risks associated with this work must we strive to avoid?

Medical Diagnosis

Diagnosing disease will likely be one of the first places where deep learning makes its mark. In just the last few years, models have been published that match or exceed the accuracy of expert humans at diagnosing many important diseases. Examples include pneumonia, skin cancer, diabetic retinopathy, age-related macular degeneration, heart arrhythmia, breast cancer, and more. That list is expected to grow very rapidly.

Many of these models are based on image data: X-rays, MRIs, microscope images, etc. This makes sense. Deep learning's first great successes were in the field of computer vision, and years of research have produced sophisticated architectures for analyzing image data. Applying those architectures to medical images is obvious low-hanging fruit. But not all of the applications are image-based. Any data that can be represented in numeric form is a valid input for deep models: electrocardiograms, blood chemistry panels, DNA sequences, gene expression profiles, vital signs, and much more.

In many cases, the biggest challenge will be creating the datasets, not designing the architectures. Training a deep model requires lots of consistent, cleanly labeled data. If you want to diagnose cancer from microscope images, you need lots of images from patients both with and without cancer, labeled to indicate which are which. If you want to diagnose it from gene expression, you need lots of labeled gene expression profiles. The same is true for every disease you hope to diagnose, for every type of data you hope to diagnose it from.

Currently, many of those datasets don't exist. And even when appropriate datasets do exist, they are often smaller than we would like. The data may be noisy, collected from many sources with systematic differences between them. Many of the labels may be inaccurate. The data may only exist in a human-readable form, not one that is easily machine-readable: for example, free-form text written by doctors into patients' medical records.

Progress in using deep learning for medical diagnosis will depend on creating better datasets. In some cases, that will mean assembling and curating existing data. In other cases, it will mean collecting new data that is designed from the start to be suitable for machine learning. The latter approach will often produce better results, but it also is much more expensive.

Unfortunately, creating those datasets could easily be disastrous for patient privacy. Medical records contain some of our most sensitive, most intimate information. If you were diagnosed with a disease, would you want your employer to know? Your neighbors? Your credit card company? What about advertisers who would see it as an opportunity to sell you health-related products?

Privacy concerns are especially acute for genome sequences, because they have a unique property: they are shared between relatives. Your parent, your child, your sibling each share 50% of your DNA. It is impossible to give away one person's sequence without also giving away lots of information about all their relatives. It is also impossible to anonymize this data. Your DNA sequence identifies you far more precisely than your name or your fingerprint. Figuring out how to get the benefits of genetic data without destroying privacy will be a huge challenge.

Consider the factors that make data most useful for machine learning. First, of course, there should be lots of it. You want as much data as you can get. It should be clean, detailed, and precisely labeled. It should also be easily available. Lots of researchers will want to use it for training lots of models. And it should be easy to cross reference against other datasets so you can combine lots of data together. If DNA sequences and gene expression profiles and medical history are each individually useful, think how much more you can do when you have all of them for the same patient!

Now consider the factors that make data most prone to abuse. We don't need to list them, because we just did. The factors that make data useful are exactly the same as the ones that make it easy to abuse. Balancing these two concerns will be a major challenge in the coming years.

Personalized Medicine

The next step beyond diagnosing an illness is deciding how to treat it. Traditionally this has been done in a "one size fits all" manner: a drug is recommended for a disease if it helps some reasonable fraction of patients with that diagnosis while not producing too many side effects. Your doctor might first ask if you have any known allergies, but that is about the limit of personalization.

This ignores all the complexities of biology. Every person is unique. A drug might be effective in some people, but not in others. It might produce severe side effects in some people, but not in others. Some people might have enzymes that break the drug down very quickly, and thus require a large dose, while others might need a much smaller dose.

Diagnoses are only very rough descriptions. When a doctor declares that a patient has diabetes or cancer, that can mean many different things. In fact, every cancer is unique, a different person's cells with a different set of mutations that have caused them to become cancerous. A treatment that works for one might not work for another.

Personalized medicine is an attempt to go beyond this. It tries to take into account every patient's unique genetics and biochemistry to select the best treatment for that particular person, the one that will produce the greatest benefit with the fewest side effects. In principle, this could lead to a dramatic improvement in the quality of healthcare.

If personalized medicine achieves its potential, computers will play a central role. It requires analyzing huge volumes of data, far more than a human could process, to predict how each possible treatment will interact with a patient's unique biology and disease condition. Deep learning excels at that kind of problem.

As we discussed in Chapter 10, interpretability and explainability are critical for this application. When the computer outputs a diagnosis and recommends a treatment, the doctor needs a way to double check those results and decide whether or not to trust them. The model must explain why it arrived at its conclusion, presenting the evidence in a way the doctor can understand and verify.

Unfortunately, the volumes of data involved and the complexity of biological systems will eventually overwhelm the ability of any human to understand the explanations. If a model "explains" that a patient's unique combination of mutations to 17 genes will make a particular treatment effective for them, no doctor can realistically be expected to double-check that. This creates practical, legal, and ethical issues that will need to be addressed. When is it right for a doctor to prescribe a treatment without understanding why it's recommended? When is it right for them to ignore the computer's recommendation and prescribe something else? In either case, who is responsible if the prescribed treatment doesn't work or has life-threatening side effects?

The field is likely to develop through a series of stages. At first, computers will only be assistants to doctors, helping them to better understand the data. Eventually the computers will become so much better than humans at selecting treatments that it would be totally unethical for any doctor to contradict them. But that will take a long time, and there will be a long transition period. During that transition, doctors will often be tempted to trust computer models that perhaps shouldn't be trusted, and to rely on their recommendations more than is justified. As a person creating those models, you have a responsibility to consider carefully how they will be used. Think critically about what results should be given, and how those results should be presented to minimize the chance of someone misunderstanding them or putting too much weight on an unreliable result.

Pharmaceutical Development

The process of developing a new drug is hugely long and complicated. Deep learning can assist at many points in the process, some of which we have already discussed in this book.

It is also a hugely expensive process. A recent study estimated that pharmaceutical companies spend an average of $2.6 billion on research and development for every drug that gets approved. That doesn't mean it costs billions of dollars to develop a single drug, of course. It means that most drug candidates fail. For every drug that gets approved, the company spent money investigating lots of others before ultimately abandoning them.

It would be nice to say that deep learning is about to sweep in and fix all the problems, but that seems unlikely. Pharmaceutical development is simply too complicated. When a drug enters your body, it comes into contact with a hundred thousand other molecules. You need it to interact with the right one in just the right way to have the desired effect, while *not* interacting with any other molecule to produce toxicity or other unwanted side effects. It also needs to be sufficiently soluble to get into the blood, and in some cases must cross the blood–brain barrier. Then consider that once in the body, many drugs undergo chemical reactions that change them in various ways. You must consider not just the effects of the original drug, but also the effects of all products produced from it! Finally, add in requirements that it must be inexpensive to produce, have a long shelf life, be easy to administer, and so on.

Drug development is very, very hard. There are so many things to optimize for all at once. A deep learning model might help with one of them, but each one represents only a tiny part of the process.

On the other hand, you can look at this in a different way. The incredible cost of drug development means that even small improvements can have a large impact. Consider that 5% of $2.6 billion is $130 million. If deep learning can lower the cost of drug development by 5%, that will quickly add up to billions of dollars saved.

The drug development process can be thought of as a funnel, as shown in Figure 12-1. The earliest stages might involve screening tens or hundreds of thousands of compounds for desired properties. Although the number of compounds is huge, the cost of each assay is tiny. A few hundred of the most promising compounds might be selected for the much more expensive preclinical studies involving animals or cultured cells. Of those, perhaps 10 or fewer might advance to clinical trials on humans. And of those, if we are lucky, one might eventually reach the market as an approved drug. At each stage the number of candidate compounds shrinks, but the cost of each experiment grows more quickly, so most of the expense is in the later stages.

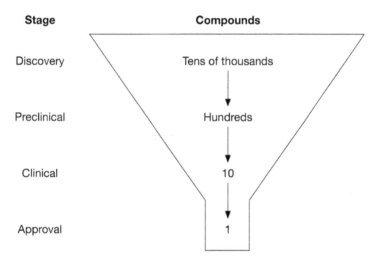

Figure 12-1. The drug development funnel.

A good strategy for reducing the cost of drug development can therefore be summarized as: "Fail sooner." If a compound will ultimately be rejected, try to filter it out in the early stages of the development process before hundreds of millions of dollars have been spent on clinical trials. Deep learning has great potential to help with this problem. If it can more accurately predict which compounds will ultimately become successful drugs, the cost savings will be enormous.

Biology Research

In addition to its medical applications, deep learning has great potential to assist basic research. Modern experimental techniques tend to be high-throughput: they produce lots of data, thousands or millions of numbers at a time. Making sense of that data is a huge challenge. Deep learning is a powerful tool for analyzing experimental data and identifying patterns in it. We have seen some examples of this, such as with genomic data and microscope images.

Another interesting possibility is that neural networks can directly serve as models of biological systems. The most prominent application of this idea is to neurobiology. After all, "neural networks" were directly inspired by neural circuits in the brain. How far does the similarity go? If you train a neural network to perform a task, does it do it in the same way that the brain performs the task?

At least in some cases, the answer turns out to be yes! This has been demonstrated for a few different brain functions, including processing visual,[1] auditory,[2] and movement sensations. In each case, a neural network was trained to perform a task. It was then compared to the corresponding brain region and found to match its behavior well. For example, particular layers in the network could be used to accurately predict the behavior of specific areas in the visual or auditory cortex.

This is rather remarkable. The models were not "designed" to match any particular brain region. In each case, the researchers simply created a generic model and trained it with gradient descent optimization to perform some function—and the solution found by the optimizer turned out to be essentially the same as the one discovered by millions of years of evolution. In fact, the neural network turned out to more closely match the brain system than other models that had been specifically designed to represent it!

To push this approach further, we will probably need to develop entirely new architectures. Convolutional networks were directly inspired by the visual cortex, so it makes sense that a CNN can serve as a model of it. But presumably there are other brain regions that work in very different ways. Perhaps this will lead to a steady back and forth between neuroscience and deep learning: discoveries about the brain will suggest useful new architectures for deep learning, and those architectures in turn can serve as models for better understanding the brain.

---

1 Yamins, Daniel L. K. et al. "Performance-Optimized Hierarchical Models Predict Neural Responses in Higher Visual Cortex." Proceedings of the National Academy of Sciences 111:8619–8624. *https://doi.org/10.1073/pnas. 1403112111.* 2014.

2 Kell, Alexander J. E. et al. "A Task-Optimized Neural Network Replicates Human Auditory Behavior, Predicts Brain Responses, and Reveals a Cortical Processing Hierarchy." *Neuron* 98:630–644. *https://doi.org/10.1016/ j.neuron.2018.03.044.* 2018.

And of course, there are other complicated systems in biology. What about the immune system? Or gene regulation? Each of these can be viewed as a "network," with a huge number of parts sending information back and forth to each other. Can deep models be used to represent these systems and better understand how they work? At present, it is still an open question.

Conclusion

Deep learning is a powerful and rapidly advancing tool. If you work in the life sciences, you need to be aware of it, because it's going to transform your field.

Equally, if you work in deep learning, the life sciences are an incredibly important domain that deserves your attention. They offer the combination of huge datasets, complex systems that traditional techniques struggle to describe, and problems that directly impact human welfare in important ways.

Whichever side you come from, we hope this book has given you the necessary background to start making important contributions in applying deep learning to the life sciences. We are at a remarkable moment in history when a set of new technologies is coming together to change the world. We are all privileged to be part of that process.

Index

Symbols

2D3U protein-ligand complex, 73
    visualization of, 74
3D protein snapshots (see proteins, structures)

A

activation functions, 10
    choices of, 11
active and decoy molecules, 178
    calculated properties of, 178-184
Adam algorithm, 14
addiction, digital, 141
adenine (A), 84
Alzheimer's disease progression, classification
    with deep learning, 138
amino acids, 61, 84
    common, chemical structures, 61
    residue, 74
angstroms, 73
antibacterial agents for gram-negative bacteria,
    111
antibody-antigen interactions, 80
    modeling antigen-antibody binding, 80
area under the curve, 31
    (see also ROC AUC scores)
aromatic rings, 68
    detection with DeepChem RdkitGridFeatur-
    izer, 70
    pi-stacking interactions, 69
artifacts of sample preparation, 115
AspuruGuzikAutoEncoder class (DeepChem),
    155
atomic featurization, 71
atomic force microscopy (AFM), 108

atoms in molecules
    chemical bonds connecting, 40
    conversion to nodes in molecular graphs, 45
autoencoders
    defined, 149
    variational, 150-151

B

bacteria
    automated systems for separating gram-
        negative and gram-positive, 119
    gram-negative, developing antibacterial
        agents for, 111
    gram-positive and gram-negative, 110
    gram-positive, mesosome in, 115
BalancingTransformer class (DeepChem), 29
baseline models, 77
bases (DNA), 84
    complementary, 94
    one-hot encoding in convolutional model
        for TF binding, 89
batches (of samples), 14
    Python function iterating over, 93
Bayesian networks
    failure to win broad adoption, 130
    probabilistic diagnoses with, 129
BBBC bioimage datasets, 117
biases, 8
binding affinities of biomolecular complexes,
    71
binding site motif, 88
biological targets, binding of molecules with, 26
biology
    knowledge of, 27

cytosine (C), 84

D

data, importance in contemporary life sciences, 2

databases in life sciences, 3

datasets
   BBBC bioimage datasets, 117
   DeepChem Dataset objects, 24
   featurization, 26
      (see also featurization)
   preparing for model prediction in virtual screening example, 189-192
   preparing for predictive modeling in virtual screening, 178
   processing and cleaning, challenges of, 29
   processing image datasets, 118
   training sets, 13
   training, validation, and test datasets used in DeepChem for molecule toxicity prediction, 27
   use in life sciences, 3
"Daylight Theory Manual", 55
dc.models.Model class, 30
dc.molnet module, 27
de Broglie wavelength, 106
decision tree classifiers, 77
decoder/encoder, 150
decoy molecules (see active and decoy molecules)
deep learning
   advances through application of, 2
   background, resources for further reading, 21
   development of antibiotics for gram-negative bacteria, 111
   for genomics, 83-97
      chromatin accessibility, 91-94
      RNA interference, 94-96
      transcription factor binding, 88-91
      unique suitability of deep learning, 87
   for medicine, 127-147
      applications in radiology, 134-141
      computer-aided diagnostics, 127-129
      deep networks vs. expert systems and Bayesian networks, 130
      diagnosing diaetic retinopathy progression, 142-145

   electronic health record (EHR) data, 130-134
   ethical considerations, 145
   job losses and, 146
   learning models as therapeutics, 141
   probabilistic diagnosis with Bayesian networks, 129
   introduction to, 7-21
      hyperparameter optimization, 17
      linear models, 8-10
      mathematical function for most problems, 7
      models, other types of, 18-20
      multilayer perceptrons, 10-12
      regularization, 15
      training models, 13-14
      validation, 15
   new ways of planning synthesis of drug-like molecules, 160
   overcoming the diffraction limit, 110
   prospects and perspectives, 201-207
      in biology research, 206
      in medical diagnosis, 201-203
      in pharmaceutical development, 204-205
   super-resolution techniques, 109
deep models, 12
   vs. shallow models, 12
deep neural networks, 1
DeepChem, 4, 23-38
   applications other than life sciences, 33
   atomic featurizer, 71
   case study using, 6
   convolutional neural networks (CNNs), 50
   Dataset objects, 24
   diabetic retinopathy model, explaining predictions of, 164-167
   featurizing a molecule, 48-49
      SMILES strings and RDKit, 48
   implementing cell counting in, 117-119
   implementing cell segmentation in, 121-124
   implementing diabetic retinopathy convolutional network in, 143
   implementing VAE model to generate new molecules, 155
   MultitaskRegressor model, 76
   RdkitGridFeaturizer, 66
      building for PDBBind featurization, 75
      implementation details, 70
   reasons for using, 23

failure to win broad adoption, 130
limitations of, 128
explainability, 174, 203
ExponentialDecay (DeepChem), 155
extended-connectivity fingerprints (ECFPs), 48
in grid featuriation, 70

F

Fast Healthcare Interoperability Resources (FIHR) specification, 132
featurization, 26
biophysical, 65-71
atomic featurization, 71
grid featurization, 66-71
featurizing a molecule, 47-49
extended-connectivity fingerprints (ECFPs), 48
learning from the data, 50
molecular descriptors, 49
featurizing molecules, 193
featurizing the PDBBind dataset, 75-79
building RdkitGridFeaturizer, 75
GraphConv DeepChem featurizer, 200
molecular, 39
using DeepChem to convert SMILES strings to graph convolution, 52
filters for interfering molecules in biological assays, 189
fixation of biological samples for microscopy, 111
fixative agents, 112
fluorescence microscopy, 113-115
deep learning and, 124
fluorescent tagging, 114
fluorophores, 113
fully connected layers, 19, 30, 34, 35, 118
functional super-resolution microscopy, 109
functions
basic function for many deep learning problems, 7
design in machine learning, 8
in multilayer perceptrons, 10
universal approximator, 12
fundus images, 142

G

GANs (see generative adversarial networks)
gated recurrent unit (GRU), 20

generative adversarial networks (GANs), 151-152
adversarial aspect, 152
evolution of, 154
structure of, 151
vs. VAEs, 152
generative models, 6, 149-161
applications in life sciences, 152-154
future of generative modeling, 154
protein design, 153
tool for scientific discovery, 154
generating new ideas for lead compounds, 153
generating new molecules, key issues with, 159
generative adversarial networks (GANs), 151-152
variational autoencoders, 149-151
working with, 155
analyzing output, 156
generator (in GANs), 151
genes, 85
disabling, using RNA interference, 94
splice variants, 86
genetics, 4
vs. genomics, 83
genome, 83
genomics, 5, 83-97
chromatin accessibility, 91-94
deep learning for
optimizing inputs for TF binding model, 168-170
DNA, RNA, and proteins, 83-85
how genomes really work, 85-87
privacy concerns for datasets, 202
RNA interference, 94-96
transcription factor binding, 88-91
vs. genetics, 83
geometric configurations, protein bonding and, 70
GPUs
advantages for running deep learning workloads, 38
high-resolution images and, 143
processing cell counting DeepChem dataset, 119
gradient descent algorithm, 13
graph convolutional networks, 50
graph convolutions, 49

About the Authors

Bharath Ramsundar is the cofounder and CTO of Datamined, a blockchain company enabling the construction of large biological datasets. Datamined aims to generate the datasets needed to accelerate the ongoing boom of AI in biotech. Bharath is also the lead developer and creator of DeepChem.io, an open source package founded on TensorFlow that aims to democratize the use of deep learning in drug discovery, and the cocreator of the MoleculeNet benchmark suite.

Bharath received a BA and BS from UC Berkeley in EECS and mathematics and was valedictorian of his graduating class in mathematics. He recently finished his PhD in computer science at Stanford University (all but the dissertation) with the Pande group, supported by a Hertz Fellowship, the most selective graduate fellowship in the sciences.

Peter Eastman works in the bioengineering department at Stanford University developing software for biologists and chemists. He is the lead author of OpenMM, a toolkit for high-performance molecular dynamics simulation, and is a core developer of DeepChem, a package for deep learning in chemistry, biology, and materials science. He has been a professional software engineer since 2000, including serving as VP of engineering for Silicon Genetics, a bioinformatics software company. Peter's current research interests include a focus on the intersection between physics and deep learning.

Pat Walters heads the Computation and Informatics group at Relay Therapeutics in Cambridge, MA. His group focuses on novel applications of computational methods that integrate computer simulations and experimental data to provide insights that drive drug discovery programs. Prior to joining Relay, he spent more than 20 years at Vertex Pharmaceuticals, where he was Global Head of Modeling and Informatics.

Pat is a member of the editorial advisory board for the *Journal of Medicinal Chemistry*, and previously held similar roles with *Molecular Informatics* and *Letters in Drug Design & Discovery*. He continues to play an active role in the scientific community. Pat was the chair of the 2017 Gordon Conference on Computer-Aided Drug Design and has been instrumental in a number of community-driven efforts to evaluate computational methods, including the NIH-funded Drug Design Data Resource (D3R) and the American Chemical Society TDT initiative. Pat received his PhD in organic chemistry from the University of Arizona, where he studied the application of artificial intelligence in conformational analysis. Prior to obtaining his PhD, he worked at Varian Instruments as both a chemist and a software developer. Pat received his BS in chemistry from the University of California, Santa Barbara.

Vijay Pande, PhD, is a general partner at Andreessen Horowitz, where he leads the firm's investments in companies at the cross section of biology and computer science, including the application of computation, machine learning, and artificial intelligence into biology and healthcare, as well as the application of novel transformative scientific advances. He is also an adjunct professor of bioengineering at Stanford, where he advises into pioneering computational methods and their application to medicine and biology, resulting in over two hundred publications, two patents, and two novel drug treatments.

As an entrepreneur, Vijay is the founder of the Folding@Home Distributed Computing Project for disease research, which pushes the boundaries of the development and application of computer science techniques (such as distributed systems, machine learning, and exotic computer architectures) into biology and medicine, in both fundamental research and the development of new therapeutics. He also cofounded Globavir Biosciences, where he translated his research advances at Stanford and Folding@Home into a successful startup, discovering cures for dengue fever and ebola. In his teens, he was the first employee at video game startup Naughty Dog Software, maker of *Crash Bandicoot*.

Colophon

The animal on the cover of *Deep Learning for the Life Sciences* is a male Sonnerat's junglefowl (*Gallus sonneratii*), also known as the gray junglefowl. The species name sonneratii is a tribute to Pierre Sonnerat, a French naturalist and explorer. Sonnerat's junglefowl are native to southern and western India, which Sonnerat visited several times between 1774 and 1781. The bird's natural habitat is forest undergrowth and bamboo thickets, but they thrive in a variety of environments, from forest to tropics to plain.

Most of the Sonnerat's junglefowl's feathers are spotted white and brown. On the tips of their wings and tails, they have black feathers with a twinge of blue. Male and female Sonnerat's junglefowl differ drastically by many measures. Roosters reach about 30 inches in length, while hens grow to only about 15 inches. The roosters are brighter than hens with glossy tails, golden speckles, and redish comb, legs, and wattle. The hens' legs are yellow and their feathers are a more dull brown.

Clutches of these chicks are born with pale brown to beige coloring. The hens typically lay 4 to 7 eggs between February and May. They lay the eggs in nests on the ground lined with grass and twigs and incubate alone, without help from the rooster.

Sonnerat's junglefowl are ancestors of domesticated chickens. They can breed with common chickens and red junglefowl (both *Gallus gallus*), creating a variety of hybrid species. When domesticated, they should be housed with solid walls around them because they are skittish. The spotted brown and white feathers are commonly used by fishermen for flies.

O'REILLY®

There's much more where this came from.

Experience books, videos, live online training courses, and more from O'Reilly and our 200+ partners—all in one place.

Learn more at oreilly.com/online-learning

Printed in the USA
CPSIA information can be obtained
at www.ICGtesting.com
JSHW050601201024
72016JS00010B/216

9 781492 039839